T0211758

Rethinking Peace and Conflict Studies

Kirsten J. Fisher
TRANSITIONAL JUSTICE FOR CHILD SOLDIERS
Accountability and Social Reconstruction in Post-Conflict Contexts

Maria Raquel Freire and Maria Grazia Galantino (*editors*)
MANAGING CRISES, MAKING PEACE
Towards a Strategic EU Vision for Security and Defense

Daria Isachenko
THE MAKING OF INFORMAL STATES
Statebuilding in Northern Cyprus and Transdniestria

Stefanie Kappler
LOCAL AGENCY AND PEACEBUILDING
EU and International Engagement in Bosnia-Herzegovina, Cyprus and South Africa

Sara McDowell and Máire Braniff
COMMEMORATION AS CONFLICT
Space, Memory and Identity in Peace Processes

SM Farid Mirbagheri
WAR AND PEACE IN ISLAM
A Critique of Islamic/ist Political Discourses

Audra L. Mitchell
LOST IN TRANSFORMATION
Violent Peace and Peaceful Conflict in Northern Ireland

Frank Möller
VISUAL PEACE
Images, Spectatorship and the Politics of Violence

Chavanne L. Peercy
LOCAL LEADERSHIP IN DEMOCRATIC TRANSITION

Michael Pugh
LIBERAL INTERNATIONALISM
The Interwar Movement for Peace in Britain

Oliver P. Richmond and Audra Mitchell (*editors*)
HYBRID FORMS OF PEACE
From Everyday Agency to Post-Liberalism

Emil Souleimanov
UNDERSTANDING ETHNOPOLITICAL CONFLICT
Karabakh, South Ossetia and Abkhazia Wars Reconsidered

Lynn M. Tesser
ETHNIC CLEANSING AND THE EUROPEAN UNION
An Interdisciplinary Approach to Security, Memory and Ethnography

Mandy Turner and Omar Shweiki (*editors*)
DECOLONISING PALESTINIAN POLITICAL ECONOMY
De-development and Beyond

Rethinking Peace and Conflict Studies
Series Standing Order ISBN 978–1–4039–9575–9 (hardback) &
978–1–4039–9576–6 (paperback)
(*outside North America only*)

You can receive future titles in this series as they are published by placing a standing order. Please contact your bookseller or, in case of difficulty, write to us at the address below with your name and address, the title of the series and one of the ISBNs quoted above.

Customer Services Department, Macmillan Distribution Ltd, Houndmills, Basingstoke, Hampshire RG21 6XS, England

---

Spatializing Peace and Conflict

Mapping the Production of Places, Sites and Scales of Violence

Edited by

Annika Björkdahl
Professor, Lund University, Sweden

Susanne Buckley-Zistel
Professor, Center for Conflict Studies, Philipps University Marburg, Germany

palgrave
macmillan

First published 2016 by
PALGRAVE MACMILLAN

The authors have asserted their rights to be identified as the authors of this
work in accordance with the Copyright, Designs and Patents Act 1988.

Palgrave Macmillan in the UK is an imprint of Macmillan Publishers Limited,
registered in England, company number 785998, of Houndmills,
Basingstoke, Hampshire, RG21 6XS.

Palgrave Macmillan in the US is a division of Nature America, Inc., One
New York Plaza, Suite 4500, New York, NY 10004-1562.

Palgrave Macmillan is the global academic imprint of the above companies
and has companies and representatives throughout the world.

PDF ISBN: 978-1-349-71553-4
E-PDF ISBN: 978-1-137-55048-4
DOI: 10.1057/9781137550484

Distribution in the UK, Europe and the rest of the world is by Palgrave
Macmillan®, a division of Macmillan Publishers Limited, registered in
England, company number 785998, of Houndmills, Basingstoke,
Hampshire RG21 6XS.

A catalog record for this book is available from the Library of Congress.

A catalogue record for the book is available from the British Library.

Contents

Part III Boundaries and Borders

Part IV Places and Sites

Figures and Tables

Figures

Tables

Acknowledgements

This volume is an outcome of an ongoing collaboration and discussion among scholars with an interest in where peace and conflict take place. Some of the chapters were presented as papers on the panel Spatial Agency in Transitions to Peace at the conference Power and Peacebuilding of the Humanitarian and Conflict Response Institute in Manchester in September 2013, where the idea for this volume was born. Others formed part of a panel at the International Studies Association conference in 2015 entitled Analysing Peace and Conflict from a Spatial Perspective.

The contributors to this volume convincingly show that spatial analysis can provide new and important insights into processes of peace and the dynamics of conflict as situated within and constitutive of different spaces and places. By mapping the interconnectedness between space and place on the one hand and peace and conflict on the other, this volume critically investigates how territorialities and scales; global and local; boundaries and borders; and places and sites are analytically employed in the study of peace and conflict. Hence we want to extend our gratitude to the authors in this volume for sharing their academic insights and knowledge, making this a novel contribution to peace and conflict studies.

We also want to thank our colleagues and friends – Stephanie Kappler at Durham University and Johanna Mannergren Selimovic at the Swedish Institute of International Affairs – who have offered insights, and stimulating and creative discussions during this project. In addition, we want to extend our thanks to Oliver Richmond at Manchester University, the editor of the Palgrave series *Rethinking Peace and Conflict Studies*, for encouragement, support and engaging discussions that have developed our ideas and helped us to advance the spatial turn in peace and conflict studies. Moreover, Friederike Mieth and Robert Nagel deserve special gratitude for their copy-editing work on the manuscript. Susanne Buckley-Zistel extends her gratitude to the Käte Hamburger Kolleg *Global Cooperation Research* which offered her a fellowship in the context of which this volume emerged. Annika Björkdahl wishes to express her gratitude to the Swedish Development Agency SIDA and the Swedish Research Council for providing research funding which made this volume possible.

Contributors

Annika Björkdahl is Professor of Political Science at Lund University, Sweden. Her research includes international and local peacebuilding with a particular focus on urban peacebuilding, and gender and transitional justice. Her recent publications include *Rethinking Peacebuilding: The Quest for Just Peace in the Middle East and the Western Balkans* (2013), *Divided Cities: Governing Diversity* (2015) and she is the editor of *Cooperation and Conflict*.

Susanne Buckley-Zistel is Professor of Peace Conflict Studies and Director of the Center for Conflict Studies, Philipps University Marburg, Germany, and is also senior fellow at the Käte Hamburger Kolleg for Global Cooperation Research at the University of Duisburg-Essen, Germany. She holds a PhD in international relations from the London School of Economics, UK, and has held positions at King's College, London, UK, the Peace Research Institute Frankfurt, Germany, and the Free University, Berlin, Germany. Her research focuses on issues pertaining to peace and conflict, violence, gender and transitional justice.

Karen Büscher is assistant professor and postdoctoral researcher in the Conflict Research Group, Ghent University, Belgium. Her research includes urbanization in the specific context of violent conflict, focusing on central Africa (DRC and Uganda), and is based mostly on ethnographic research. Several pieces of her work have appeared in international, peer-reviewed journals.

Sven Chojnacki is Professor of Comparative Politics and Peace/Conflict Research at the Free University Berlin, Germany, and head of the research project Variances and Consequences of Territorial Control by Non-State Actors at the Collaborative Research Center (SFB) 700 at the university. He is currently doing research on the spatial dynamics of collective violence from the perspective of conflict studies, critical geography and post-colonial theory.

Jolle Demmers is associate professor and co-founder of the Centre for Conflict Studies at Utrecht University, the Netherlands. She lectures and writes on theories of violent conflict, ethnographies of neoliberalism and diaspora and violent conflict. Her latest book, *Theories of Violent Conflict* (2012), has been nominated for the ENMISA 2013 Distinguished Book Award.

Martin Doevenspeck is Professor of Geographical Conflict Research and Political Geography at the University of Bayreuth, Germany. His research focuses on migration, spatial dimensions of violent conflict and the political dimensions of climate change and risk in West and Central Africa.

Faye Donnelly is a lecturer at the University of St Andrews, UK. She is author of *Securitization and the Iraq War: The Rules of Engagement in World Politics* (2013). Her current research focuses on security emblems, the interrelationships between security and borders, and the intersections between securitization and finance.

Bettina Engels is Junior Professor of Conflict Research, focusing on Sub-Saharan Africa, at the Otto Suhr Institute for Political Science, Free University Berlin, Germany. Her research focuses on conflict over land and mining, spatial and action theory, and resistance, urban protest and social movements in Africa.

Nina Fischer teaches Jewish Studies at the University of Edinburgh, UK. Previously, she held fellowships at the Australian National University and the Hebrew University of Jerusalem, Israel, and served as researcher and project manager of the History and Memory Group at the University of Konstanz, Germany. Her research areas include memory, holocaust and Middle Eastern studies, and she is currently working on a book tracing the conflicting cultural representation of Jerusalem. She is author of *Memory Work: The Second Generation* (2015).

Andreas Hackl is a PhD candidate in social anthropology at the University of Edinburgh, UK, writing on Israel-Palestine and Palestinians in Israel in particular. Previously he was a Jerusalem-based news correspondent and independent analyst for various media outlets and the humanitarian news agency of the United Nations (UN). A graduate of social anthropology and political sciences at the University of Vienna, Austria, he is primarily a political anthropologist with particular research expertise in issues of identity, civil resistance and mobility. He is also a board member of Peace and Conflict Studies in Anthropology, a network of the European Association of Social Anthropologists.

Annika Henrizi is a PhD candidate and occasional lecturer at the Center for Conflict Studies, University of Marburg, Germany. In her PhD project, Gendered Agency in (Post-)Conflict Spaces: Women's Engagement in Iraqi NGOs (non-governmental organizations), she analyses the engagement of Iraqi women within NGOs in the wider context of peacebuilding. Her research is located at the nexus between sociology, and peace and conflict studies. Her special focus is on theories of space and action. She is also working on gender in conflict and post-conflict societies, and sociopolitical processes in the Middle East.

Kristine Höglund is Professor in the Department of Peace and Conflict Research, Uppsala University, Sweden. Her research has covered issues such as the causes and consequences of electoral violence, the importance of trust in peace-negotiation processes and the role of international actors in dealing with crises in war-torn societies. Her work has been published in journals such as *Democratization, Review of International Studies, Negotiation Journal, International Negotiation* and *International Peacekeeping*.

Milena Komarova is a research fellow in the Institute for the Study of Conflict Transformation and Social Justice, Queen's University Belfast, UK. Her research interests include ethnonational urban conflicts and conflict transformation; public space, place and collective identities; and mundane mobilities and urban borders. She has published on discourses of peacebuilding, urban borders in post-socialist cities, 'shared space', territoriality and regeneration in Belfast, and the uses of visual and mobile methodologies in conflict research.

Erik Melander is Professor at the Department of Peace and Conflict Research and, since 2006, Deputy Director of the Uppsala Conflict Data Program (UCDP), Uppsala University, Sweden. He has also been a senior research associate at the Joan B Kroc Institute for International Peace Studies, University of Notre Dame, France, since 2007. His current main interests are third-party peacemaking in civil wars and the relationship between gender equality and armed conflict.

Laura Michael is a postdoctoral research fellow at the Queen's University Belfast, UK. Her research focus is on sacralization, space and the planning system, and she has drawn on a range of conflict and post-conflict cases to understand the interplay between public policy, place-making and memorialization.

Brendan Murtagh is a chartered town planner and a reader in the School of Planning Architecture and Civil Engineering at Queen's University Belfast, UK. He has researched and written widely on urban regeneration, conflict and community participation. His recent books include *The Politics of Territory, Segregation, Violence and the City* and *Understanding the Social Economy and the Third Sector*.

Henri Myrttinen is the head of the Gender Team at International Alert. Since 2000 he has been researching, working, teaching, training and publishing on gender, especially in the context of post-conflict and conflict-affected societies in Central and Eastern Europe, Sub-Saharan Africa and Southeast Asia for a variety of NGOs, think-tanks and donor agencies. He holds a PhD in conflict resolution and peace studies from the University of KwaZulu-Natal, South Africa.

Liam O'Dowd is Professor of Sociology and Director of the Centre for International Borders Research at Queen's University, Belfast, UK. His research interests include urban conflicts in contested states, borders, imperialism and nationalism, the Northern Ireland conflict and the sociology of intellectuals. He has published extensively on cities, borders and Northern Ireland, and most recently has co-authored/edited *Religion, Violence and Cities* (2014).

Linda Price has researched widely on rural sociology with a particular emphasis on meaning and identity in the context of agricultural restructuring. She has extensive experience in the application of ethnographic methods and life histories to reveal the complexity of gender and farming cultures, traditions of rural communities, and impacts of change on farming men and women across the life course. This work has been undertaken in the UK and Ireland, and more recently in Canada and Australia.

Olivera Simić is a senior lecturer with the Griffith Law School, Griffith University, Australia, and a visiting professor with the UN University for Peace in Costa Rica. Her research engages with transitional justice, international law, gender and crime from an interdisciplinary perspective. She has published numerous journal articles and book chapters, two monographs and four edited collections in the field of transitional justice. Her latest publications are *The Arts of Transitional Justice: Culture, Activism, and Memory after Atrocity* (with Peter D Rush, 2013) and *Surviving Peace: A Political Memoir* (2014).

Margareta Sollenberg is an assistant professor in the Department of Peace and Conflict Research, Uppsala University, Sweden. Her research has covered economic incentives for armed conflict, specifically with regard to the role of foreign aid, and various issues relating to conflict data collection. She has been involved in the UCDP for the past two decades and has published on UCDP data in, for example, *Journal of Peace Research, SIPRI Yearbook* and *Conflict, Security & Development*.

Elena B. Stavrevska is a PhD candidate in political science at the Central European University, Hungary. Her work focuses on the conceptualization of agency, in particular local agency, in (post-)conflict societies. During her doctoral studies she has been a visiting fellow at the University of Toronto, Canada, and the Institut für Europäische Politik in Berlin, Germany. She was also part of the team working on the Seventh Framework Programme project Cultures of Governance and Conflict Resolution in Europe and India. She has done extensive ethnographic research across Bosnia and Herzegovina (BiH) and currently works with the Institute for Research and Social Innovation in Skopje, Macedonia.

Ralph Sundberg holds a PhD and is a researcher in the Department for Peace and Conflict Research, Uppsala University, Sweden. He has also been a project manager within the UCDP since 2008. His current main interests are the subnational study of civil war and the social psychology of violence.

Mikel Venhovens holds an MA degree in conflict studies and human rights from Utrecht University, the Netherlands. He has widely researched violent conflict in Eastern Europe and the former Soviet Union, with a distinct focus on the concept of governmentality in relation to spatiality (national) identity formation and separatist conflicts. He has conducted field research in Latvia, Georgia and Abkhazia.

Zala Volčič is a senior lecturer and researcher at the Centre for Critical and Cultural Studies, University of Queensland, Australia. In addition to her current research, which deals with the role of the media in the (commercial) construction of national identities, other projects include writings on media memories, public TV, social movements and media education. She has published five books and over 60 articles and book chapters on the role of the media in the Balkans, and she has edited a collection on transitional justice in the former Yugoslavia.

Spatializing Peace and Conflict: An Introduction

Annika Björkdahl and Susanne Buckley-Zistel

Introduction

This edited volume is a response to a *cri de coeur* for research to investigate the interconnectedness between space, peace and conflict. It calls for a careful rethinking of their relationship in order to understand where peace and conflict 'take place'. Spatial analysis can provide new and important insights into the dynamics of conflict and processes of peace as situated within and constitutive of different spaces and agencies. Material structures such as borders and boundaries, war zones, dividing bridges and peace gardens determine how various agents manoeuvre and how they encounter each other. At the same time, spatial features are the result of these forms of interaction, and social interactions may lead to material expressions, reshaping and changing spaces, and as a consequence the way individuals and groups can move within them. There is thus a duality of space and agency; they are mutually constitutive.

Advancing the spatial turn we currently observe in international relations and politics, this volume suggests that spatial theory provides analytical concepts which have not yet been fully exploited in analysing peace and conflict. To do so it raises the question of how we can understand peace and violent conflict in spatial terms. The authors set out to critically investigate how space, scale and sites can be conceptualized and analytically employed in the field, proposing a new research agenda that advances a spatial approach to the study of conflict and peace. Thus the volume provides novel theoretical frameworks, reconceptualization of key concepts, in-depth case studies and a cross-case comparative analysis that explore peace and conflict through spatial approaches.

The relational combined with the spatial is at the core of an understanding of sites of peace and conflict. By adopting a transcalar approach it seems possible to explore the relationship between conflict and peace, and space and place, and how they interplay across different scales. We understand peace and conflict located in place to be about a sociospatial relation that is always made and remade. It means that peace and conflict become connected to the spatial. To understand peace and conflict as situated makes it possible to reconceptualize and contextualize them.

We often assume we can localize where peace and war take place. We understand peace to be the opposite of war, and thus we assume that where war is present, peace is absent. Peace emerges when conflict abates. In that sense, it seems difficult to move beyond the binaries of war and peace as both negative peace (defined as the absence of direct violence) and positive peace (defined as the absence of direct as well as structural and cultural violence) relate to violence and perhaps must do so (Galtung 1969). Yet war and peace are clearly intertwined, and if there ever was a clear line between them it has become increasingly blurred. Foucault, for example, argues that war is inside peace, while Mac Ginty describes the continuities between war and peace as a situation of no war – no peace (Foucault 2003; Mac Ginty 2006). For in the midst of violent conflict there are islands of peace, and in times of peace there are outbreaks of violent conflict. By situating peace and war in time and space we are better able to grasp them as fleeting notions and to comprehend how they can coexist.

Since this is the first edited volume that analyses conflict and peace processes through the lens of space, the approach taken is deliberately comprehensive in order to provide insights into the many forms that spatial analysis can assume. Bringing together a range of scholars, it offers an interdisciplinary synthesis of their spatial dimensions. Contributors have academic backgrounds in critical geography, peace and conflict studies, critical anthropology, political science, cultural studies, sociology and urban studies. They employ methodologies ranging from large-N analysis to ethnography. Yet despite their different interdisciplinary approaches, the authors share core theoretical assumptions concerning the significance of spatiality: all chapters employ space as an analytic category and develop strong theoretical contributions, as well as offering new empirical insights based on original research. Drawing on concepts such as spatial governmentality, scalar politics, relational spatial theory and spatial narratives, they investigate case studies from cities such as Belfast, Dili and Jerusalem, via sites such as rape camps and karaoke bars, to countries such as the Democratic Republic of the Congo (DRC) or Bosnia and Herzegovina (BiH).

The spatial turn

The central contention of this volume is that the organization of space is significant for the structure and function of peace as well as war. This perspective is, however, by no means a new phenomenon and it has given rise to an increased interest in the notion of space in social science, geography, anthropology, history and other academic disciplines. In the late 1960s, Michel Foucault observed that while the 19th century had a strong focus on history, the 20th could be described as the epoch of space (Foucault 1967/1986: 331). In a similar vein, Frederic Jameson remarked that 'A certain spatial turn has often seemed to offer one of the more productive ways of distinguishing postmodernism from modernism proper' (1991: 154), again drawing attention to a shift from merely historicizing social relations to considering their spatial situatedness as well.

But what does a spatial turn signify? The expression was first used by Edward W. Soja (1989) in his monograph *Postmodern Geographies* when referring to a shift in geography and the social sciences from considering space as a purely material condition to understanding it as a social product. In this sense he situates spatial analysis in the history of 'turns', most prominently the linguistic turn. Soja draws on the work of the French sociologist Henri Lefebvre according to whom space is always social because it assigns more or less appropriate locations to social relations (Lefebvre 2009: 186–187). It is only through space that abstract social relations become concrete. Importantly, though, space itself is socially produced, it is a result of interactions and it can thus be described as a complex social construction composed of social norms, values and ascribed meanings (Lefebvre 1991: 26). From this it follows that space is both a complex social construction and the condition in which individuals and groups interact. In a circular way, space provides the structures that enable and constrain agency. There is thus a mutually constitutive relationship between space and the societies that inhabit it: space is shaped by social interactions and at the same time it shapes these interactions.

The latter aspect resonates in the work of the critical geographer Doreen Massey for whom 'space is... part (a necessary part) of the generation, the production, of the new. In other words the issue here is not to stress only the production of space but space itself as integral to the production of society' (Massey 1999: 10). However, this does not imply that material space in and of itself has certain properties which render this possible, but rather that it is in social practice that these properties emerge. In this sense, as Chojnacki and Engels (Chapter 1) illustrate, the production of space as introduced by Lefebvre opens up the view on how the material and the social are related to each other.

For Massey, spaces are not containers but rather permeable structures. They are not dehistoricized, largely homogenous, fixed and bounded entities in which interaction occurs but rather the materialization of social relations which have sedimented over time and which are therefore contingent (Massey 2007: 154). Moreover, the spatial combination of social relations which constitute the uniqueness of any locality is not confined to this place but stretches beyond its (permeable) boundaries so that the inside defines part of the outside, and the outside part of the inside (ibid.: 5).

The production of space – for Lefebvre and others – is highly political: 'There is a politics of space because space is political' (Lefebvre 2009: 33). For if space is produced socially, it is always a contested terrain, reflecting some position while excluding others. Interest and strategies collide and are determined by prevailing power asymmetries, potentially causing conflicts. Space must therefore not simply be understood as concrete and material but simultaneously as ideological and subjective (Warf and Arias 2009: 3). Every society produces its own space (Lefebvre 1991), and this can be distinguished between perceived, conceive and lived space (Soja 1996). Such space is produced through acts of naming, as well as the distinctive activities and narratives associated with particular social and material spaces.

Spatial theory has long attempted to disentangle the relationship between space and place. Such efforts include an understanding of how spaces are transformed into places by agents through spatial practices and the spatialization of social activities. By spatialization, social anthropologist Setha Low (1996, 2000) refers to the practice of locating physically and conceptually social relations and social practice in social and material space. Thus the social *production* of spaces reveals the material aspects of creating place. The social *construction* of space unmasks the immaterial aspect of creating place – that is, the actual transformation of space through people's social exchanges, memories, images and daily use of the material place. Both processes are social in the sense that both production and construction of space are contested for political reasons.

A number of categories emerge in the context of spatial theory and will thus be explored briefly. First, sites can be described as more or less bounded types of space, a lived space, a place, defined by and constructed in terms of the lived experience of people. As such they express a sense of belonging for those living at a site. Being 'in place' involves a range of cognitive (mental) and physical (corporeal) performances that are constantly evolving as people encounter the site. Thus conflict and peace are performed through everyday practices at sites such as a border-line, memorials, cityscapes or trenches. Scales, second, can be defined as operational and methodological in the sense that analytical complexities such as patterns, processes and agents can be located according to their scale of operation. Thus scales are (re)produced through social and political processes of conflict or peace, and they affect the operation of these processes. Agency, lastly, refers to the human capacity to act – a capacity that is not exercised in a vacuum but in a social world in which structure shapes the opportunities and resources available in a constant interplay of practices and narratives (Giddens 1984; see also Cleaver 2007). In order to understand what entices structures such as the material and social legacy of conflict to change requires an understanding of how differently positioned agents can take part in conflict or peace.

Space in peace and conflict studies

Analysing violent conflicts and peace processes from a spatial perspective is slowly but steadily becoming part of peace and conflict studies. So far there has been no sustained inquiry into the relationship linking peace and conflict with space, despite the fact that this analytic perspective is essential for the field. Yet what is the value of a spatial turn for peace and conflict studies?

First, it draws attention to the fact that spatial features are already very much at the centre of peace and conflict, as well as peace and conflict studies. War zones, border disputes, fights over land and territory, out-of-area missions, inter- or intrastate conflict, besieged cities, peace gardens and war memorials, and many other terminologies make reference to spatial features. This points to the fact that the conduct of violence and the maintenance of peace always have a spatial element to them.

In order to include spatial features in their analysis, a number of scholarly contributions draw on or emerge from the discipline of geography (Gregory 2010; Koopman 2011; McConnell et al. 2014; Megoran 2011). Historically, war and geography have been strongly interrelated. To date, thinking in territorial terms has been widespread in military and security circles. In recent years, though, scholars have started to critically reflect on the links between space and violence, moving from a classical geopolitical understanding towards a multidimensional perspective on space and violent conflict. For instance, Gregory's work reveals the multiple ways in which contemporary wars are delinked from territory, including the 'War on Terror', and how they are socially constructed and reach into the lives of ordinary people (Gregory 2004, 2010, 2011). Ó Tuathail and Dalby (1998) critically examine the geopolitical condition of international politics that challenges the way we think about state borders, power, territory defence and security, and they explore the spatialization of boundaries and dangers that construct geopolitical representations of self and other (Ó Tuathail 2000). In a similar vein, Gregory (1994) provides an insightful analysis of spatial imaginaries of 'otherness' that problematizes the distinctions between self and other. Following on from this, feminist geographers explore the spatial construction of gender roles in times of war and point to the differential spaces ideologically constructed for men and women as a consequence of nationalism and war (Dowler 2005).

While earlier contributions to peace studies, such as by Starr (1991) and Kirby and Ward (1986), work with notions of space as fixed entities and units of analysis, more recent studies emphasize the social construction of spaces and territories. For instance, drawing on critical geopolitics, human geography and social theory, Higate and Henry (2009, 2010) analyse United Nations (UN) peacekeeping as a territorializing practice, which transforms spaces and thus shapes the perception of everyday security in conflict societies. Against the backdrop of peacekeeping in Haiti, Kosovo and Liberia, they illustrate how this is performed by particular security practices. The production of territoriality, and the strategies and power relations this entails, is moreover addressed by Doevenspeck (Chapter 2).

Second, using space as an analytic category exposes the fact that both war and spatial structures are products of human activities (Flint 2004: 5). People establish borders, define territories, develop war zones, and construct social, political and economic topographies, which are then fought over or form elements in a peace process. There is thus an intrinsic link between the production of space through agency and the occurrence of violent conflict or peace. According to Thrift (2003: 95), space 'is the outcome of a series of highly problematic temporary settlements that divided and connect things up into different kinds of collectives which are slowly provided with the means which render them durable and sustainable'. Thus, space has importance for understanding conflict, and its dynamics and legacy, as well as for sustainable peace and durable coexistence in post-conflict countries (cf. Kobayashi 2009; McConnell et al. 2014).

Against this backdrop, recent spatial approaches to peace and conflict studies question the very dichotomy between war and peace. For Flint (2011: 31), situating war and peace in time and space discloses the fact that they are intrinsically intertwined manifestations of dynamic social processes and cannot be treated separately. Their connection stems from the fact that many social relations within a particular context contribute to the construction of both war and peace. In this sense, 'War is the means for the construction of a new complex of social relations and peace is a complex of social relations that produces winners and losers, and hence domination and conflict' (ibid.). Flint thus concludes that negotiation, continual political tension and, at times, open conflict form part of social situations that are generally referred to as war and peace, and that it is more useful for their analysis to think of them 'as a continuum of social settings in which power relations and political differences always exist, but in varying degrees of consensus and non-violence compared to disagreement and violence' (ibid.: 45). Instead of differentiating between war and peace, the social construction of post-conflict spaces should be considered as an embodiment of power relations because these spaces both reflect prevailing power relations and produce and reproduce new ones. This resonates with the argument of Büscher (Chapter 4).

Third, the analysis of scales is also of relevance to peace and conflict studies. Recognizing that peace and conflict dynamics can operate at multiple scales and that they can also be, and often are, transcalar processes because they intersect with multiple processes operating on any other scale, scale like space is an outcome of social activities and processes (Swyngedouw 1997). Much research on conflict and peace contains implicit or explicit ideas about their locus, and recent studies that draw on new data from the Uppsala Conflict Data Program (UCDP) explore the spatial dimensions of conflict by making use of disaggregated and geocoded data from Sub-Saharan Africa (Sunberg and Melander et al. 2013; see also Höglund, Melander, Sollenberg and Sundberg, Chapter 3). By using this spatially disaggregated information, the authors are able to move beyond the state level as the unit of analysis and study the exact location of violent events.

The way scalar levels are referred to by various audiences might help to understand the politics behind the different readings. Note, for instance, (topological) markers such as global, global North, centre or Western on one side and local, global South, periphery and local-local on the other, often used as synonyms within their ascribed containers, albeit referring to profoundly different social and political constellations. One topographic binary that is often deployed in peace and conflict studies is between local and global (or international) where, at least according to more conventional approaches, interventions in war, but also peacebuilding, take place. This is sometimes referred to as top down or bottom up, suggesting that there are discrete units which can be analysed and which impact on each other, often with the local being portrayed as the victim of the global. Thus the scalar levels of global and local draw attention to the social construction of space and agency over time in a manner that unsettles the boundaries between the two. Thrift (2004: 59) argues that a 'scale-dependent notion' of space assumes a nested hierarchy of scales from 'global' to 'local' and needs to be 'replaced by an emphasis on connectivity'. To further problematize the interplay between the scalar levels of global and local in a way that unsettles their boundaries, ideas about frictional encounters between the global and the local in peacebuilding have been advanced, and to understand the outcome of such encounters, research on hybrid peace, hybrid political orders and hybrid governance has broken new grounds (Björkdahl et al. 2016; Boege et al. 2009; Jarstad and Belloni 2012; Mac Ginty 2010).

In a similar vein, Stefanie Kappler questions this binary view of the global and the local, arguing that 'peacebuilding agency and the identity categories that actors create for themselves are fluid, transversal and movable, with actors constantly resituating themselves between competing forces of identification' (Kappler 2015: 876). Peacebuilding should hence be considered as a multilayered process in which identities are constantly shifting and being renegotiated along the line of the political, economic and social aspects of the actors involved. Susanne Buckley-Zistel (2016), too, stresses the mutual constitution of global and local as scales. Drawing on Massey, she argues that they cannot be considered as binaries but that they are related concepts. As a consequence, the local is not merely the victim of the global but always also has the potential for agency – that is, to contribute to the shaping of the global. This potentiality opens up room for imagining new social and political spatialities. Henrizi (Chapter 6) provides an illuminating case study in this regard.

Fourth, the multidimensional effects of particular places and sites on conflict-affected societies, too, can be studied by using space as an analytic category. Indeed, certain sites and places appear to offer symbolic, political and social frameworks or capacities which maintain the social, political, identity-oriented and governance structures of a conflict while at other times supporting peacebuilding efforts. Through a spatial perspective we can explore how places and sites are negotiated and how they shape people's experiences, memories, feelings and interpretations. This might include memorials, museums and parks which commemorate past violence. Often they form contested terrains and reflect the different currents and views in a society, which might be expressed in their architectural and thus symbolic language. Regarding the remembrance of violence, it has been argued that 'Analysing the architectural language of memorials opens an avenue for explaining the various aesthetic, political, and ethic concepts and intentions, as well as didactic approaches underpinning contemporary memorial projects' (Buckley-Zistel and Schäfer 2014: 15). This offers an analytical framework to assess the complexities of sites of commemoration. Simić and Volčič (Chapter 14) reflect this reasoning concerning acknowledgement remembrance in relation to sites of crime.

Certain places foreground the politics of belonging (Yuval-Davis 2006). These are also discursive sites for the construction of narratives in which experiences are rendered meaningful, and some of the researchers in this volume unpack such narrative meanings and trace their embodiment in material space. This is, for instance, central to Fischer's contribution (Chapter 12) which argues that the spatial narrative of the Arab past and present in Jerusalem – reflected in the Dome of the Rock – is an articulation of belonging and a demonstration of identifying with this contested space. Recent research by Björkdahl and Mannegren (forthcoming) zooming in on bridges in the post-war Bosnian landscape reveals these structures as flawed metaphors for connecting people and depicts them as key sites for agonistic peacebuilding as well as contestation. Moreover, drawing on Foucault's notion of heterotopia, Buckley-Zistel (2014) analyses how the former Stasi remand prison Hohenschönhausen in Berlin serves as a space of compensation for former inmates and activists, while Gough (2000) assesses various landscape types of sites of remembrance, such as peace parks, which increasingly replace war memorials.

Fifth, borders and boundaries are commonly defined as the lines dividing distinct political, social and/or legal territories, often provoking conflict about their demarcation. Borders result from processes of bordering that differentiate between places, people and jurisdictions (Diener and Hagen 2012). Bordering processes thus highlight borders as a force and resource in international and domestic political, social and economic relations. Such a multifaceted approach also sheds light on the contingency and variability in bordering practices across both space and time. As this volume reveals, what constitute borders is contested academically, conceptually, as well as empirically and politically. Research on borders and boundaries includes interrogations of various 'walls', such as between the two Germanies, but also the 'peace walls' in Belfast and the 'security barrier' between Israel and Palestine. It may also cover immaterial borders, such as mental and cognitive barriers that constrain mobility, and construct stereotypes and prejudice. Such walls are symbolic and material manifestations of political boundaries. Demmers and Venhovens (Chapter 8) illustrate the spatialities of contested statehood in a way that resonates with the spatial approach to borders and boundaries.

Borders and boundaries are also a key concern in the evolving research on urban space and divided cities (Bollens 2012; Pullan 2011). As material and symbolic assets, cities become contested spaces in many conflicts. They are also more likely to become the locus of identity-based conflicts than to transcend such conflicts as multiple identities converge in cities, and national institutions are increasingly incapable of mediating or negotiating diversity and difference when mutually exclusive identities become the basis of politics (Bollens 2012). Divided cities play host to diversity and difference, and they are susceptible to intense intercommunal, often identity-based, conflict and violence reflecting exclusive identities, and ethnic or nationalist fractures, and they have proved resistant to peacebuilding efforts. As Hackl (Chapter 9), Fischer (Chapter 12) and Komarova and O'Dowd (Chapter 13) reveal, cities challenged by violent conflict hold actors who attempt to assert control over spaces and places and over actions and interactions in ways that become part of the daily practice of territoriality in the cityscape. For example, in identity-based conflicts, identities are secured and ensured by ascribing them into the cityscape, and when identities become merged with territory, neighbourhoods become ethnicized.

About this volume

The chapters in this volume all use space as an analytic category. Based on existing contributions to spatial theory – as well as developing new theoretical perspectives – they investigate notions such as territoriality, urban spaces, global–local encounters, spatial governmentality, borders, sites, scalar politics and spatial narratives. The research methodologies employed reach from quantitative methods of data analysis of novel and disaggregated data from the UCDP dataset to qualitative case studies based on narrative analysis, fieldwork, participant observations, interviews and various methods of inquiry to collect historically, politically and personally situated accounts, descriptions and representations of human lives. They provide a point of departure for comprehensive, empirical analysis relating to peace and war, and a way to explore the various discourses expressed through spatial practice and specific spatial elements of societies in conflict of societies in transition between war and peace and of post-conflict societies.

Part I focuses on the notions of territory and scale in the spatial analysis of peace and conflict. Chojnacki and Engels begin (Chapter 1) by offering a critical review of conflict studies literature that focuses on spatial categories. Assessing contributions which analyse inter- and intrastate violence and war, they depict how, since the 1970s, geographic markers such as scale, borders, natural resources, territory and infrastructure have been deployed to explain the occurrence of violence. Moreover, they review literature which draws on environmental conflicts through ecological change as ways of explaining war and violence. The authors take a critical perspective when they highlight the fact that much of this literature is based on the notion of the state as the site where intra- and interstate violence takes place. In doing so, they argue, conflict analysts perpetuate the assumption that the state is a pregiven, natural entity and the central unit for analysing and assessing violent conflicts. Drawing on Lefebvre's notion of the social production of space, the authors propose an alternative understanding of space; they suggest considering conflict as a form of social action that is constitutively related to space. In this reading, space does not determine conflict but rather space and conflict mutually produce and reproduce each other. The authors thus stress the interplay between the social and political production of space on the one hand and conflict action on the other.

This ties in with Doevenspeck's contribution (Chapter 2), which assumes the perspective of political geography to analyse successful and unsuccessful modes of producing territoriality by rebel movements in the DRC. Producing territoriality, he argues, is crucial in shaping the outcome of violent conflict, rendering it an important, albeit often overseen, aspect for their analysis. This is particularly relevant regarding the establishment of political order by non-state violent actors in times of civil war and unrest. Based on fieldwork in North Kivu among the politicomilitary movements CNDP (Congrès National pour la Défense du Peuple/National Congress for the Defence of the People) and M23 (March 23 Movement), the author outlines the geography of violence which is based on an imagined enclave of a delicate local peace and a predictable political order in the ongoing civil war in the east of the DRC. For the author, space is not simply a unit in which conflict manifests itself but the basis of the analytical concepts of territory and territoriality. He illustrates the cultural and social production of territories, including the power structures they entail, drawing attention to the political dimensions of space and the spatial dimension of politics. The author shows how multiple technologies, practices and discourses serve to shape a particular space with a territorial shape (border, boundaries, etc.), a symbolic shape (flags, statues, parades, etc.) and an institutional shape (administration, politics, education, etc.). Considering territory as an effect, rather than a material entity, he concludes, draws the focus of analysis to its modes of production and is thus of significant conceptual value.

A slightly different approach is taken by Höglund, Melander, Sollenberg and Sundberg (Chapter 3), who offer a quantitative analysis of the scalar dimension of violent conflict, war and unrest. They examine whether different forms of violence occur predominantly in rural or urban environments, if and how the phenomena have the same or different origins and dynamics, and if and how they are linked to each other. They draw on geocoded and disaggregated data about violent conflicts that occurred between 1989 and 2010 in Sub-Saharan Africa, and which have been compiled by the UCDP. Three insights can be gleaned from their analysis. The authors first show that large-scale violence is rare and that it takes place in few countries and few areas. All cases that experienced large-scale violence did so in both urban and rural environments, suggesting that there is a link between these scales, and they conclude that the analysis of violence in such large-scale cases should thus not be limited to either rural or urban environments. A second insight of the authors' large-N analysis suggests that violent conflict occurs more often in urban than in rural areas. This, they contend, may be due to higher population densities as well as the symbolic and economic relevance of urban centres. However, their third insight is that there are a comparatively greater number of casualties in rural areas – that is, even though rural violence is less frequent it may have more of an impact on the population. In addition, despite the fact that violence often occurs in urban areas, its origins may stem from structural violence and marginalization outside cities.

Part II scrutinizes the constitution of the local and the global, and how they interact with each other. Focusing on an urban area of violent strife, Büscher (Chapter 4) provides an ethnographic analysis and spatial reading of the transformation of Goma, a city in the East of the DRC, which has experienced violence over the past 20 years, turning it into a conglomerate of local and global agents. She draws attention to the interaction between the production and reproduction of the physical urban topography and the spatial practices of conflict and peace by illustrating how the city has turned into a booming political and economic centre. Drawing on rich empirical material, the author illustrates how in the urban landscape both local effects of civil war and local and global dynamics of peacebuilding are apparent. Goma is often referred to as the 'city of rebellion', the 'city of refuge' and the 'UN fortress', but different actors, such as warlords, businessmen and UN staff, strongly impact on its governance, emerging social networks, livelihoods and opportunities, as well as the material landscape. This is *inter alia* visible in the ever-changing use and function of various spaces, such as the transformation of public spaces into military camps or humanitarian offices, and of refugee camps into residential areas. Moreover, it manifests itself in the increasing contestation between urban military elites over particular neighbourhoods and streets. The author's analysis also focuses on the agency of 'big men', and on internally displaced persons camps and international peace missions, but also on the tourism industry that develops on the back of this. All of these are significantly transforming the city over time. Both local and international actors thus contribute to the production of the city space in this urban 'no war-no peace' zone.

For Myrttinen (Chapter 5), too, the encounter of global and local dynamics in a time between war and peace in an urban space is crucial. He portrays the East Timorese capital Dili during its years of crises, 2006–2008, which were marked by violence and political instability. He does so through an intersectional perspective combining gender, class, age and racialized dynamics, and thus offers a spatial analysis of the way various groups that live in the same urban space seem little connected in person, although their modes of existence are. Drawing on extensive field research in Dili, the author focuses on four separate yet related spaces in which local and global players interact. First he portrays internally displaced persons camps where many East Timorese escaping violence in their country spent a considerable amount of time. The street forms his second area of analysis and illustrates how violence between police and peacemakers on one side and martial arts groups, gangs and ritual arts groups on the other erupts. This is followed by a depiction of the hotel as a site in which international peacekeepers and other expatriates exist in a world that seems far removed from the one outside its doors. Lastly, the karaoke bar is examined as a site where Indonesian, Filipino, Malaysian and Chinese businessmen and labourers, as well as local hustlers and sexworkers, congregate for recreation or work. By illustrating these four urban spaces, the author reveals how they are interrelated despite their apparent secludedness, and how the dynamics of peacebuilding manifest themselves in an urban space.

Continuing with the theme of local and global spaces, Henrizi (Chapter 6) draws on relational space theory to challenge the relationship between the scales. In contrast with the view that local and global are discrete containers which stand in binary opposition to each other, a relational reading of space suggests that there is a duality of acting and space so that space is conditioned by agency, and vice versa. Hence scales such as global and local cannot be discrete entities but always have the potential to interrelate and overlap. The author's empirical focus is on the implementation of the Convention on the Elimination of All Forms of Discrimination against Women (CEDAW) in post-war Iraq. In the process of localizing CEDAW, different actors such as Iraqi women's non-governmental organizations (NGOs), international women's NGOs, and representatives from the government and the UN negotiate its implementation and the transferral of the norms it entails. Drawing on her field research data, the author shows how local Iraqi NGOs operate within a global structure – a globally constituted space – in order to not only promote their agenda but also contribute to the very production of the global space. Her case study hence confirms that the scalar levels of global and local do not exist independent of each other but are mutually constitutive. Neither global nor local can be considered as static, but they change over time. Moreover, she points to the link between space and agency – for instance, illustrated by the possibility of local Iraqi NGOs entering the UN building in Geneva and positioning themselves and their concerns which they felt was a great achievement – as well as to the importance of power in the process of renegotiating local and global.

Part III draws attention to borders and boundaries, and how they interact with and are mediated by agents of peace and conflict. Such structures stress the functional character of spatial elements, such as territorial and state borders, frontiers, walls and barriers, and their effect on cooperation and coordination. Borders are regimes of governance in which control is exercised through the management of space (Merry 2001). Inspired by the notion of spatial governmentality, Stavrevska (Chapter 7) critically investigates the spatialization of ethnicity in BiH. Her analysis demonstrates the imagining of ethnicity as possessing spatial characteristics and as such being the most powerful political category, superior to the state, which is imagined simply as a conglomerate of the ethnicities of the three constitutive peoples of BiH. By focusing on the everyday, the analysis sheds light on the importance of class structures and socioeconomic commonalities in overcoming the legacy of the conflict, and produce spaces of peace and solidarity. In this way, the author's findings challenge conventional wisdom, which holds that class structure as a dimension of life often fuels conflict rather than creating experiential conditions to overcome ethnic division and conflict. Here the spatial analysis adds a valuable component to the analysis of post-conflict societies by focusing on the spaces that people produce, occupy and interpret. Based on extensive ethnographic fieldwork balancing an insider–outsider perspective, the author provides original empirical material and analysis while taking into account her own positionality. She has collected various voices and everyday experiences of people through semistructured interviews, revealing that the spatialization of ethnicity has been widely accepted by the governed and has been the basis of ethnic spatial governmentality.

Employing ideas connected to ethnographies of the state and state spatiality, Demmers and Venhovens (Chapter 8) examine contested statehood and the making and unmaking of borders by the parties to the Abkhazian–Georgian conflict. They investigate the spatial dimensions of the mechanisms of government deployed by actors in conflict and the ways in which they imagine the state. When 'reading war' through border landscapes and border architectures, the authors find multiple spatialities at work. Their analysis, which builds on empirical insights gained from fieldwork in Abkhazia and Georgia, shows how authorities from both Georgia and Abkhazia claim space and produce subjectivities in highly contested ways through architectural design, security devices and multiple bureaucratic rituals and routines. The authors show that borders and landscapes are produced and contested, and how claims to space are materialized in border security posts, checkpoints and barbed wire. Post-conflict landscapes are thus never stable or fixed but contested and renegotiated, and through these processes we can learn about the shifting topographies of power and the dynamics of conflict.

Hackl (Chapter 9) rescales to the urban and investigates urban space as an agent of conflict and 'peace'. He explores the mundane struggles of Palestinians in Tel Aviv in order to provide a more comprehensive understanding of how relationships between marginal urban subjects and a dominant urban space simultaneously shape projects of cooperation and conflict. The author argues that space executes an agency that is both abstract and relational, and that shifting subjectivities relate to such space in various ways. As the spatial engagements of Palestinians in Tel Aviv have remained largely unexplored as Tel Aviv itself performs an identity within which the Palestinians seemingly do not exist, the author fills a gap in the urban studies literature on minorities in majority cityscapes. He draws on extensive anthropological fieldwork in Tel Aviv, mapping spaces of cooperation and conflict in the everyday of Palestinian workers, Palestinian women seeking employment, Palestinian citizens of Israel in the high-tech sector and Palestinian protesters. An important finding is that the agency that space can exercise draws on the power differentials that operate in space and are produced by it, and thus it has an impact on the everyday emergence, recurrence and dissolution of peace and conflict.

Donnelly (Chapter 10) uncovers new spatial stories of (de)securitization of the Olympic Games. Her analysis reframes the Olympics as a space of contestation rather than a place of peace. In so doing it attempts to disentangle the interrelationship between space and place. Through de Certeau's concept of spatial stories, the analysis focuses on how actors attempt to impose order on fluid spaces and draws empirical insights from the 2014 Sochi Winter Olympics. It reveals how the Olympics becomes a space in which securitization can occur. An important finding of this chapter is that space is constituted by, and constitutive of, multidimensional spheres of interaction. The spatial stories told about and enacted in the Olympics have ramifications, and it is becoming a product of spatial stories of securitization. However, empirical findings challenge the securitization stories, suggesting that the Russian Government and the International Olympic Committee sought to continuously desecuritize the Sochi Winter Olympics to reaffirm it as a place of peace. Through its focus on transforming spaces into places, the chapter then bridges to the final part of the volume.

Part IV moves us from space to place. A spatial perspective on places and sites clearly reveals the interrelatedness of the concepts space and place as 'place can be seen as a portion of space in which people dwell together' and a 'locality' or a site where things happen or matters take place (Agnew 2011). Michael, Murtagh and Price (Chapter 11) explore sites of memorialization and sacralization in the post-conflict space of Belfast. Through an in-depth analysis of the Maze Long Kesh (MLK) Prison, they demonstrate how memorialization and sacralization are constitutive of place-making and how place is inextricably linked to memory, memorialization and identity. Furthermore, they evaluate the confrontation of public policy with places loaded with meaning and memory for ethnic groups determined to legitimate their past as well as their future claims to space. The authors show how space is socially constructed and how memoralization elevates place from the mundane to the sacred. In this respect, heritage becomes a political resource to bolster identity and a connection to a cause, especially as with MLK Prison,[1] where it is concretized in a material site. Such a site reveals where the conflict is territorialized and where space becomes both a material and a non-material resource to be claimed. It thus does not comply with the reimagining of Northern Ireland as a liberal, economically progressive and socially readjusted place.

Fischer (Chapter 12) maintains the urban scale of analysis and provides an analysis of spatial narratives of belonging in Jerusalem centring on the Dome of the Rock – a cultural icon symbolizing the religious, national and ethnic belonging of Palestinians in Jerusalem. From a cultural studies perspective, she demonstrates the conflation of memory, belonging and politics in the contested city of Jerusalem and observes the petrification and extension of conflict in the visual symbols employed in the urban space of the city. What is revealed is how memory always tends towards spatialization and that memory is emplaced. Memorial and heritage sites are annexed in order to promote particular interpretations of the past, rendering competing spatial narratives visible. Through 'being in the place', the author adopts a reflexive methodology that includes collecting empirical material from reading the city's material and symbolic images, interpreting the narratives and actions that relate to the Dome of the Rock as well as experiential knowledge. She uses pictorial representations of the city, revealing its microcosm of conflicts. These images of urban sites are used to make claims about ownership and belonging. Jerusalem more than any other place has a complex set of stories, which are complicated by religious beliefs being turned into spatial memory and political argument. Conflict and segregation are entrenched in the cityscape but also in the visual representations of Jerusalem.

Komarova and O'Dowd (Chapter 13) explore the emerging narrative of post-conflict Belfast as a 'shared city', according to which urban places are given a new, more inclusive identity after years of violence in Northern Ireland. Drawing on sociological literature about the relationship between the social and the physical shaping of cities, as well as between spatial form, urban experience and materiality, the authors develop the methodological tool of 'spatial narratives' in order to analyse the relationship between spatial change and conflict transformation. Narratives, in this sense, are spatial stories which go beyond language to include materialities and performances. They constitute place through their grounding of social identity and through fixing contingent meaning to a place. The chapter is based on extensive fieldwork in Belfast. It illustrates how the narrative of a 'shared city' seeks to build accommodation in the present and to provide a positive outlook for the future. The authors conclude that while the 'shared city' narrative is discursively appealing since it promises conflict transformation and reconciliation after the Northern Ireland Troubles, it also entails a number of contradictory meanings. Moreover, they argue that its performative function is relatively weak and restricted to urban central spaces. Key features of the narrative include rendering communal events more inclusive, and removing signs that might be interpreted as divisive, but this has so far not been successful in turning new, alternative places into meaningful, welcoming and relevant spaces.

Simić and Volčič (Chapter 14) read two specific sites of crime – the spa hotel Vilina Vlas in Visegrad and the Court of BiH in Sarajevo – to explore the complexity of acknowledgement, remembering and forgetting, asking what role such sites have in community-based education and healing. The selected spaces are everyday ordinary sites that were reworked for and transformed by extraordinary activities, such as rape, torture and massacre, turning them into spaces of crime. The authors argue that the identity of spaces is very much connected to the spatial narratives which are recounted about them, how those stories are narrated and by whom, and to what extent they challenge or uphold the hegemonic interpretations of past events. They draw on extensive fieldwork in BiH's landscape where the legacies of conflict are spatialized. The chapter provides a narrative analysis that shows how commemoration of these places has become part of an ongoing contestation around the interpretation of the Bosnian War. The analysis also finds that national identity is constituted and reconstituted in spatial and gendered processes. As such the authors argue that space and place are central organizing terms of the national identity and for the recognition of the spatial.

In sum, all of the contributions draw implicit or explicit attention to the performative practice of agency in constituting space or when operating within space. They move beyond seeing space as a container with a clear demarcation of inside/outside and point to the constant negotiation and reshaping of these boundaries. In doing so they challenge conventional, territorial foci, such as on the state, the region or the globe. They point instead to the metaphorical meaning of places and sites such as bridges, sites of atrocity or memorials in and through which meaning is constituted. Furthermore, a number of original and novel themes can be distilled from these analyses, including social and political practices and narratives, the transformative capacity of agency, and situated processional analysis of the dynamics of peace and conflict over time. Different perspectives on space, place and site can be distinguished, such as space as a meeting point for frictional encounters of the 'global' and the 'local', sites serving as contested spaces for memory-making and practices, the metaphorical meaning of material spaces, and the interplay between social and material spaces. Moreover, the authors recognize that plural forms of conflict and peace may coexist in a given space and situate processional analysis of the dynamics of peace and conflict over time.

This volume hence provides multidisciplinary perspectives on the complex architecture of peace and war and by doing so advances the spatial turn in peace and conflict studies. What emerges strongly across the chapters is how important spatial practices are for the interpretation of peace and peacebuilding processes, as well as for war and conflict dynamics. What looks like peace from one perspective, scale or site is much more problematic if one changes perspective, jumps scale or moves to a different site. This suggests the need for a far more diverse consideration of the dimensions and geographies of peace and war. With this volume, space and spatial perspectives are more fully returned to the analysis of the topic and we therefore hope that it stimulates further discussion and research on spatializing peace and conflict.

Note

1. Unionists, in general, refer to the site as 'the Maze', the name given to the prison by the British Government – officially, Her Majesty's Prison Maze. Republicans use 'Long Kesh' because it was an internment (without trial) camp set up in the early 1970s as a response to the violence that broke out in 1969, before special category status for inmates as political prisoners was withdrawn in 1976.

References

Agnew J (2011) Space and Place. In: Agnew J and Livingstone D (eds) *Handbook of Geographical Knowledge*. London: Sage, pp. 316–331.

Björkdahl A, Höglund K, Millar G, Verkoren W and van der Lijn J (eds) (2016) Introduction: Peacebuilding through the lens of friction, in *Friction and Peacebuilding: Global and Local Encounters in Post-Conflict Societies*. Abingdon: Routledge, pp. 1–16.

Björkdahl A and Mannegren Selimovic J (2016) A Tale of Three Bridges: Agency and Agonism in Peacebuilding, *Third World Quarterly*. DOI: 10.1080/01436597.2015.1108825.

Boege V et al. (2009) On Hybrid Political Orders and Emerging States. What Is Failing: States in the Global South or Research and Politics in the West? *Berghof Handbook Dialogue Series* 8: 15–35.

Bollens SA (2012) *City and Soul in Divided Societies*. London: Routledge.

Buckley-Zistel S (2014) Detained in Hohenschönhausen. Heretrotopia, Narratives and Transitions Form Stasi Past in Germany. In: Buckley-Zistel S and Schäfer S (eds) *Memorials in Times of Transition*. Antwerp: Intersentia, pp. 97–124.

Buckley-Zistel S (2016) Frictional Spaces: Transitional Justice between the Global and the Local. In: Björkdahl A, Höglund K, Millar G, Verkoren W and van der Lijn J (eds) *Friction and Peacebuilding: Global and Local Encounters in Post-Conflict Societies*. Abingdon: Routledge, pp. 17–31.

Buckley-Zistel S and Schäfer S (2014) Introduction: Memorials in Times of Transitional. In: Buckley-Zistel S and Schäfer S (eds) *Memorials in Times of Transition*. Antwerp: Intersentia, pp. 1–28.

Cleaver F (2007) Understanding Agency in Collective Action. *Journal of Human Development* 8(2): 223–244.

Diener A and Hagen J (2012) *Borders: A Very Short Introduction*. New York: Oxford University Press.

Dowler L (2005) Amazonian Landscapes: Gender, War, and Historical Repetition. In: Flint C (ed.) *The Geography of War and Peace: From Death Camps to Diplomats*. Oxford: Oxford University Press, pp. 133–148.

Flint C (2004) Introduction. In: Flint C (ed.) *The Geography of War and Peace: From Death Camps to Diplomats*. Oxford: Oxford University Press, pp. 1–15.

Flint C (2011) Intertwined Spaces of Peace and War: The Perpetual Dynamism of Geopolitical Landscapes. In: Kirsch S and Flint C (eds) *Reconstructing Conflict: Integrating War and Post-War Geographies*. Farnham: Ashgate, pp. 31–48.

Foucault M (1967/1986) Of Other Spaces: Utopias and Heterotopias. In: Leach N (ed.) *Rethinking Architecture: A Reader in Cultural History*. Abingdon: Routledge, pp. 330–336.

Foucault M (2003) *'Society Must Be Defended': Lectures at the College de France, 1975–76*. New York: Picador.

Galtung J (1969) Violence, Peace, and Peace Research. *Journal of Peace Research* 6(3): 167–191.

Giddens A (1984) *The Constitution of Society: Outline of the Theory of Structuration*. Cambridge: Polity Press.

Gough P (2000) From Heroes' Groves to Parks of Peace: Landscapes of Remembrance, Protest and Peace. *Landscape Research* 25(2): 213–228.

Gregory D (1994) *Geographical Imaginations*. Oxford: Blackwell.

Gregory D (2004) *The Colonial Present: Afghanistan, Palestine, Iraq*. Malden, Oxford, Carleton: Blackwell.

Gregory D (2010) War and Peace. *Transactions of the Institute of British Geographers* 35: 154–186.

Gregory D (2011) The Everywhere War. *The Geographical Journal* 177(3): 238–250.

Higate P and Henry M (2009) *Insecure Spaces: Peacekeeping, Power and Performance in Haiti, Kosovo and Liberia*. London: Zed Books.

Higate P and Henry M (2010) Space, Performance and Everyday Security in the Peacekeeping Context. *International Peacekeeping* 17(1): 32–48.

Jameson Fredric (1991) *Postmodernism, or, the Cultural Logic of Late Capitalism.* Duke University Press: Durham.

Jarstad A and Belloni R (2012) Introducing Hybrid Peace Governance: Impact and Prospects of Liberal Peacebuilding. *Global Governance: A Review of Multilateralism and International Organizations* 18(1): 1–6.

Kappler S (2015) The Dynamic Local: Delocalisation and (Re-)Localisation in the Search for Peacebuilding Identity. *Third World Quarterly* 36(5): 875–889.

Kirby A and Ward MD (1986) *The Spatial Analysis of Peace and War.* Working Paper, Program on Political and Economic Change, Institute of Behavioral Science, University of Colorado.

Kobayashi A (2009) Geographies of Peace and Armed Conflict: Introduction. *Annals of the Association of American Geographers* 99(5): 819–826.

Koopman S (2011) Let's Take Peace to Pieces. *Political Geography* 30(4): 193–194.

Lefebvre H (1991) *The Production of Space.* Blackwell: Oxford.

Lefebvre H (2009) Reflections on the Politics of Space. In: Brenner N and Elden S (eds) *State, Space, World: Selected Essays/Henri Lefebvre.* Minneapolis: University of Minnesota Press, pp. 167–184.

Low S (1996) Spatializing Culture: The Social Production and Social Construction of Public Space in Costa Rica. *American Ethnologist* 23(4): 861–879.

Low S (2000) *On the Plaza: The Politics of Public Space and Culture.* Austin: University of Texas Press.

Mac Ginty R (2003) *No War, No Peace: The Rejuvenation of Stalled Peace Processes and Peace Accords.* Basingstoke: Palgrave.

Mac Ginty R (2010) Hybrid Peace: The Interaction between Top-Down and Bottom-Up Peace. *Security Dialogue* 41(4): 391–412.

Massey D (1999) Philosophy and Politics of Spatiality: Some Considerations. *Geographische Zeitschrift* 87(1): 1–12.

Massey D (2007) *Place, Space and Gender.* Cambridge: Polity.

McConnell F, Megoran N and Williams P (eds) (2014) *The Geographies of Peace.* New York: I.B. Tauris.

Megoran N (2011) War *and* Peace? An Agenda for Peace Research and Practice in Geography. *Political Geography* 30(4): 178–189.

Merry SE (2001) Spatial Governmentality and the New Urban Social Order: Controlling Gender Violence through Law. *American Anthropologist* 103(1): 16–29.

Ó Tuathail G (2000) The Postmodern Geopolitical Condition: States, Statecraft, and Security into the Twenty-First Century. *Annals of the Association of American Geographers* 90(1): 166–178.

Ó Tuathail G and Dalby S (eds) (1998) *The Geopolitics Reader.* London and New York: University of Minnesota Press and Routledge.

Pullan, W (2011) Frontier Urbanism: The Periphery at the Centre of Contested Cities. *Journal of Architecture* 16(1): 15–35.

Soja EW (1989) *Postmodern Geographies: The Reassertion of Space in Critical Social Theory.* New York: Verso.

Soja EW (1996) *Thirdspace: Journeys to Los Angeles and Other Real-and-Imagined Places.* Cambridge: Blackwell.

Starr H (1991) Joining Political and Geographic Perspectives: Geopolitics and International Relations. *International Interactions* 17(1): 1–9.

Sundberg R and Melander E (2013) Introducing the UCDP Georeferenced Event Dataset. *Journal of Peace Research* 50(4): 523–532.

Swyngedouw E (1997) Neither Global nor Local: Glocalization and the Politics of Scale. In: Cox K (ed.) *Spaces of Globalization: Reasserting the Power of the Local.* New York: Guilford Press, pp. 137–166.

Thrift N (2003) Space: The Fundamental Stuff of Human Geography. In: SL Hollaway, SP Rice and G Valentine (eds) *Key Concept in Geography.* London: Sage, pp. 95–107.

Thrift N (2004) Intensities of Feeling: Towards a Spatial Politics of Affect. *Geografiska Annaler: Series B Human Geography* 86(1): 57–78.

Warf B and Arias S (2009) The Reinsertion of Space in Social Sciences and Humanities. In: Warf B and Arias S (eds) *The Spatial Turn: Interdisciplinary Perspectives.* Abingdon: Routledge, pp. 1–10.

Yuval-Davis N (2006) Belonging and the Politics of Belonging. *Patterns of Prejudice* 40(3): 197–214.

Part I
Territorialities and Scales

1

Overcoming the Material/Social Divide: Conflict Studies from the Perspective of Spatial Theory

Sven Chojnacki and Bettina Engels

Introduction

In conflict studies it has become a trend to link space and conflict theoretically as well as empirically (see e.g. Raleigh et al. 2010; Stephenne et al. 2009; Wucherpfennig et al. 2011). So far, however, few substantial arguments have been made that consider how the relationship between space and conflict can enhance our understanding of conflicts. One challenge is that the landscape of theoretical narratives and empirical studies is rather fragmented along the line of different disciplines. Social geographers, political scientists, conflict researchers and development scholars alike suggest that we broaden our understanding of space in order to reflect on the conditions, dynamics and effects of conflict more precisely, without much engagement between these disciplines. One central question discussed in these different bodies of literature is whether space is an external, material condition that influences human action, or whether a different understanding of space is required.

In this chapter we critically review the corresponding debates in conflict studies and examine how they conceptualize spatial categories, both theoretically and empirically. We begin with an overview of how spatial references are made in the mainstream debates in conflict studies, then introduce theoretical reflections of the concept of space in social geography. Next we link the two debates in order to outline how spatial categories can be applied in the analysis of violent conflict in a more productive way. Although the main objective of the chapter is to introduce the subject and to present the state of the art, it concludes by offering a perspective on conflict as social action that is constitutively related to space, and suggesting that space does not determine conflict but that space and conflict mutually produce and reproduce each other.

In order to unfold this argument the chapter proceeds as follows. First, we review central debates within conflict studies with regard to how they refer to spatial concepts. We begin with studies on international and civil wars that, since the 1970s, have used geographic categories – such as topography, infrastructure, borders or proximity – as explanatory factors for the occurrence of armed conflict. In the study of civil war, natural resources and territorial control are particularly considered. Next the debate on environmental conflicts – that is, research on the relationship between ecological change and violent conflict – is sketched out as another influential strand within conflict studies, which refers strongly to spatial aspects. Following this we show that most studies of international and civil war and environmental conflict reproduce the container concept of the territorially bounded nation-state.

The main problem we identify in conflict studies literature is thus that most studies do not connect physical and material with sociopolitical aspects of collective conflict, therefore perpetuating what has been referred to as methodological nationalism. Wimmer and Glick Schiller (2002) define methodological nationalism as 'the assumption that the nation/state/society is the natural social and political form of the modern world' (Wimmer and Glick Schiller 2002: 302). While their concern is with outlining how the epistemic programmes and structures of mainstream social science research are interwoven with modern (capitalist) nation-state formation, in this chapter we confine our critique to one of their three variants of methodological nationalism: the focus on the nation-state as the key, and often only, unit of analysis.

After the literature review, we show how concepts of spatial theory developed in radical geography help to redress these problems. The notions of territorialization and scale are taken as examples to elucidate this in more detail. We conclude by arguing that spatial theory provides helpful analytical categories for conflict studies, but that space does not simply present an additional variable to be added to the analysis of violent conflict. The point is, rather, to examine how conflict can be analysed from the perspective of a spatial theory, and how spatial and conflict theory can enter into a productive dialogue.

Space in the analysis of international and civil wars

As early as the 1970s, research on international war conceptualized geographic conditions (e.g. direct neighbourhood and spatial distance) as explanatory factors for the occurrence of armed conflict and war (see Diehl 1991; Starr 1991). Ever since, physical features such as borders or contiguity have been identified as enabling interaction between territorial entities, or as providing opportunity structures under which actors formulate preferences and make decisions (Siverson and Starr 1991). Quantitative studies on the causes of interstate wars emphasize the importance of territorial aspects (notably boundary disputes) as the most conflicting among all contentious issues (Vasquez 2009). Here, both the understanding of geography and the twin concept of opportunity and willingness (Most and Starr 1980) are oriented towards the relatively static notion of boundaries as central to methodological nationalism.

At present, most quantitative empirical studies of the field operate in a territorial container, though, when focusing on correlations between physical variables such as topography, infrastructure, borders or proximity and conflict. In other words, most studies are based on an understanding of geography that restricts space to allegedly fixed material factors, such as the availability of resources or physical demarcation (bordering) in order to explain and predict armed conflict. One central argument holds that the dynamics of armed conflict are conditioned by the location of and distance to political, natural or other resources – for instance, when arguing that in strong regimes, civil wars are located further away from the capital (Buhaug 2010), or that conflicts last longer if they are located along remote international borders, in regions with valuable minerals or at a distance from the main government (Buhaug et al. 2009).

Other studies show that the existence, concentration and type of natural resources impacts on the occurrence of civil war. In this reading, diamonds and oil have highly significant effects, while agricultural goods are hardly significant at all (Fearon 2005; Lujala et al. 2005; Ross 2004). Moreover, centralized resources, such as petroleum or easily accessible mines, are considerably easier to monitor and protect than geographically dispersed resources, such as opium plantations, alluvial diamonds or tropical forests (Le Billon 2001). One critical aspect is also the proximity of key resources to a fighting faction's headquarters or the capital. Buhaug and Rød (2006) demonstrate furthermore that armed conflict correlates with the spatial distribution of features such as relative road density, while O'Loughlin et al. (2011) found evidence that the proximity to strategic locations (military installations, administrative institutions) affects the incidences and diffusion patterns of violence over time.

Most of these studies have in common that they rarely question the relationship between conflict and external resources as a material condition of conflict. But material features only become resources through social relations, and it is these social relations that explain why a certain resource enables or disables certain conflict action. Even though places and territories have locatable and measurable properties, the meaning and impact such conditions have for individuals and social groups are contingent on time and space. The relationship of space and conflict hence revolves around the intangible and dynamic qualities that are attributed to these conditions by social groups or individuals (e.g. notions of ownership, ideas of cultural identity[1] that are connected to places and territories). These social and political meanings also have to be reflected in approaches to conflict studies, which draw on topography, borders and resources.

Studies on civil war link conflict to the importance of territory for maintaining order, emphasizing the significance of territorial control for the political and strategic relations between violent actors and the civilian population. Using a microlevel approach, Kalyvas (2006, 2012) concludes that the level of territorial control by armed actors, as well as their desire to minimize information asymmetries, allows for predicting the spatial variation of violence against civilians in civil wars. Depending on the degree of control (reaching from areas fully controlled by one actor to areas contested by armed groups), violence becomes a function of different zones of territorial control. Kalyvas thus offers a pioneering perspective on the interconnection of information, territorial control and types of violent action. Still, his model remains within the logic of territory as a physical condition and fails to consider its cultural meanings or its political and social power relations.

Control over territory is a precondition for the extraction of natural resources. Access to these resources is directly related to political power, as well as to the opportunity and incentive to wage war. The value of territorial control, however, varies with the economic and strategic importance of specific areas. For state actors and armed groups alike, both territories with valuable resources such as diamonds, gold or oil, and those which possess an inherent value for strategic action, such as capitals, harbours and transport routes, are more important than a piece of land in the periphery without major resources.

Environmental conflict

Another influential strand in conflict research that refers to spatial categories is the study of environmental conflicts. Since the early

1990s, a debate has emerged about whether and how environmental change influences the occurrence of violent conflict on different levels (local, national and international) (Baechler et al. 2002; Gleditsch 1997; Homer-Dixon 1999). The fourth assessment report of the Intergovernmental Panel on Climate Change, Climate Change 2007 (IPCC 2007) was circulated widely in politics, media and academia. In the following years, scholars focused on the question of how climate change effects, such as rising temperature, increasing rainfall variability and extreme weather, affect violent conflict. By now, most scholars agree that ecological change becomes relevant for collective conflict only through social and political mediation (Gleditsch 1998; Salehyan 2008), although some recent studies still confirm the argument that rainfall patterns and violent conflict correlate (Raleigh and Kniveton 2012; Theisen 2012). Remarkable progress has been made with regard to empirical evidence, as well as to the specification of social and political factors that mediate the relationship between environmental change and violent conflict (Benjaminsen and Boubacar 2009; Benjaminsen et al. 2012). Some weaknesses remain, however, in particular regarding the persisting focus on the nation-state as the most important level of analysis.

Regarding environmental conflict research, few studies explicitly refer to a theoretical concept of space. Changes in the physical-material environment, as well as violent conflict, come across in different ways across time and space. Also, violence is a socially differentiated phenomenon, meaning that at a specific location at a particular moment in time, not all people are affected by collective violence in similar ways. Likewise, the social effects of ecological change vary horizontally and vertically, as numerous studies have shown (see Adger 1999, 2006; Wisner et al. 1994). Where people live (in terms of locations and territories) affects their vulnerability, but it only constitutes one aspect among several and is related to other conditions, such as social structures.

Thus the analysis of the spatial dimension of environmental conflict that exclusively refers to physical and material features and detaches them from the social structures in which they are embedded – by presenting certain areas such as river deltas, coastal zones and savannah regions as being particularly prone to ecological change effects and conflict risks, for instance – fails to deal adequately with the complexity of ecological and social systems. However, an *a priori* denial of the relevance of any physical materiality for violent conflict falls equally short. Existing models that try to integrate environmental and sociopolitical factors often assume a linear, causal relationship between ecological change and conflict, regarding social and political institutions as intervening variables (Homer-Dixon 1999).

Persisting containers: Analytical levels and scales

Recent research on civil wars and the study of environmental conflict have attempted to overcome the container concept of the nation-state (Buhaug and Gates 2002; Buhaug and Rød 2006). Nevertheless, many studies, in one way or another, still conceptualize the state as a territorial container. Even if complex disaggregated geographical data are used, the state is simply replaced by arbitrary grid cells as the central units of analysis. Such a shift in the level of analysis has yet to fully overcome methodological nationalism without merely replacing it by 'methodological territorialism' in the sense of assuming smaller, but still territorially bounded, entities. While the construction of grid cells does not reproduce the arbitrary demarcations of nation-states, it creates new containers and equally arbitrary boundaries, not taking into account any social, political or cultural meaning of territories and places. Furthermore, by deliberately disregarding nation-state borders, grid cells analysis is unable to tackle the historical and political meaning these borders have for local conflicts.

Still, recent studies using disaggregated geographical data shed some light on the occurrence of civil wars: they show, for instance, that topographic variables such as forests and mountains affect the manner in which internal violent conflicts are conducted and how they impact on the prospect of winning a battle or a war (Buhaug and Gates 2002; Buhaug and Rød 2006; Gates 2002). The analytical problems related to nation-state centrism are, however, not resolved solely by simply 'downscaling' the level of analysis (Agnew 1994). Rather, the potentials of integrating structural and geometric vector data[2] methodologically are only fully beneficial if they are systematically linked to theoretical reflections on both space and conflict.

The analytical bias in favour of the nation-state level is reflected in qualitative as well as quantitative studies (Deligiannis 2012). Even though in recent years an increasing number of case studies on civil wars and environmental conflicts have focused on the substate (mainly local) level, they hardly investigate the interscalar relationships between the local, national, transnational and global levels. At the same time, statistical analysis that is mostly based on the aggregate level of the nation-state is rarely linked to local case studies (for an exception, see Benjaminsen and Boubacar 2009). Moreover, most quantitative empirical studies do not differentiate between the territorial and social aspects of variables such as population density and growth, resource availability and the occurrence of violent conflict (e.g. Hauge and Ellingson 1998;

Hendrix and Glaser 2007). Likewise, large-N analyses of intercommunal conflicts hardly ever make spatial references beyond the national level (Reuveny 2007: 662) or take social and political institutions on the substate level into account. The choice of variables in large-N studies tends to depend on data availability rather than on theoretical considerations, as Raleigh and Urdal (2007) admit: 'Despite its theoretical importance, we do not attempt to empirically capture resource distribution as such data are currently not available on the local level' (2007: 678).[3] To summarize, while recent work in the study of civil war tries to cope with the critique of methodological nationalism by integrating sophisticated disaggregated data, which has generated illuminating results, it does not, at least from a spatial theory perspective, provide many insights into the relationship between space and conflict.

Analysing peace and conflict from a spatial theory perspective

The review of recent debates in the field of conflict studies that refer to spatial aspects has identified two main problems. First, many studies fail to systematically link the physical/material and sociopolitical aspects of collective conflicts. Second, they perpetuate the construction of the nation-state as a territorial container. In the following, we argue that spatial theory as developed in social and political geography offers useful concepts and analytical categories to address these problems, and we illustrate this by using the categories of territorialization and scale.

One central element of social and political geography is that physical and material conditions do not determine, and consequently cannot explain, social action in a causal sense. This insight is, among others, based on the work of Soja (1989, 1996), who introduced the 'spatial turn' in the 1980s (see also Duncan et al. 2004; Warf and Arias 2009). Approaches subsumed under this term refer to the social organization of space and aim to avoid essentialist concepts of space as purely physical and material. This critique of spatial determinism does not mean neglecting the influence that material relations (in a Marxist sense) and physical/material conditions (may) have on social action. According to Lefebvre, the production of space (rather than space itself) is an inherently historical process – that is, he understands space as a social relation produced by human beings as social actors. With Marxism as a central theoretical backdrop for Lefebvre (Elden 2007; Soja 1989: 120), he aims to join Marxist political economy and historical materialism with a spatial perspective.[4] In this sense, space functions as a means of establishing and maintaining control over economic resources and people, power and domination (Lefebvre 1994: 26), with 'power' referring to the control over persons and material artefacts. Thus the spatial conditions of social action, themselves socially produced and reproduced, are an expression of power relations.

Determinism, as a theoretical term, refers to anything that is 'given' in an essentialist manner, and that determines the development of anything else. Criticizing spatial determinism therefore does not imply denying that, for instance, topography or climate may have an impact on social relations, since 'having an impact' is by no means equal to 'determining'. The critique rather targets the idea that these factors determine the social world, providing a causal explanation for social action such as conflict. It remains a core question in spatial theory, though, to what degree the materiality of space has theoretical autonomy when acknowledging that it is always socially produced. While early spatial turn debates focused on the social construction of space, recent authors suggest that we have to 'attend to the materiality of social, ethical and political life' (The Space of Democracy and the Democracy of Space Network 2009: 582; see also Featherstone 2004). Space, then, is not just an outcome but part of the explanation of social action – in the interpretative sociology sense of understanding, rather than in the positivistic sense of causality (see Massey 1984).

Though spatial categories are frequently referred to in conflict studies, often this does not imply a theoretically based analysis of space but refers to (physical) material conditions, expressed in spatial terms. Most studies focus on material features, such as geographical distances, the location of natural resources, temperature or rainfall patterns, assuming that these factors cause or trigger social action and possibly collective conflict. We argue that a more comprehensive understanding is required which not only takes into account the material and the symbolic, identity-related dimension of space, but also analyses how the two dimensions relate to one another. Based upon a relational and multidimensional understanding of space, different spatial dimensions – global and local (scale), vertical and horizontal (place, territory) – should be conceptualized as mutually constitutive and relationally intertwined (Dietz et al. 2015). However, a relational spatial perspective – in peace and conflict studies, as in other fields of research – is not an end in itself:

> The relevance of a particular spatial form – either for explaining certain social processes or for acting on them – can be measured only from the perspective of the engaged actors. Thus, in order to define

criteria for the relevance of (a specific form of) spatiality, we need to start, both in our theoretical endeavours as well as in political practice, from concrete social processes and practices rather than reifying spatial dimensions.

(Mayer 2008: 416)

Keeping in mind that in some debates within the social sciences, including conflict studies, space has somehow become a buzzword, it is important to reflect what we mean by 'space' and whether and in what way it helps us to better understand what we are actually studying. Therefore we suggest, first, to take actors and their social action and practices as the starting-point for conflict analysis and, second, to reflect upon which spatial category or categories turn out to be promising for specific research questions and aims. In spatial theory, four core analytical categories can be identified (Jessop et al. 2008; see also Brenner 2008; Dietz et al. 2015): place, scale, territory/territorialization and network. In what follows we will present some ideas about how conflicts can be analysed through the application of two of them: territorialization and scale.

Territorialization

Research on conceptualizing spatial categories in a material, deterministic way widely ignores the relevance of identity construction for violent conflict. In contrast, studies in the field of identity and conflict (ethnic conflict, intercommunal conflict, nationalism and so on) mostly do not take into account the physical materiality of the respective conflicts (see e.g. Blimes 2006; Fox 2004; Gartzke and Gleditsch 2006). Spatial theory as discussed in critical geography provides an opportunity to bridge this gap. The concept of territorialization is particularly helpful here.

While early globalization theorists (e.g. Taylor 1996) argued that territorial boundaries may lose their functionality for political authority, for Newman (2010), territory loses its relevance neither for the performance of authority and domination nor for the construction of collective identities. On the contrary, territorial references are central to identity-related inclusion and exclusion – that is, for the construction of 'self' and 'other'. Political identities often (but not always) refer to territorially defined spaces, though these are not necessarily linked to nations and states but, possibly more frequently, to other 'imagined communities', such as ethnic, indigenous and autochthonous groups. It is perfectly clear that these identity constructions play an essential role for violent conflicts. For instance, most struggles over land between local groups are at the same time conflicts about whose claims to land are legitimate, which in turn is closely linked to the social negotiation of citizenship and belonging, of inclusion and exclusion (Lund 2011).

Territorialization, following Vandergeest and Peluso (1995), can be understood as state policies of spatial-administrative organization in order to establish control over natural resources and human beings, within or beyond a state's territorial borders. Territoriality, they argue, is a central element in understanding state–society relations. Sack (1986) defines territoriality as 'the attempt by an individual or group to affect, influence, or control people, phenomena, and relationships, by delimiting and asserting control over a geographic area' (1986: 19). Territoriality thus refers to the inclusion and exclusion of people within certain geographic borders, and political rulers territorialize state power in order to achieve different goals. The enforcement of taxes and the access to valuable natural resources are pivotal, but control over potential military conscripts is also decisive (Vandergeest and Peluso 1995: 390), which links internal territorialization to military disputes within and between states. Most states further use territorialization to control citizens. Political measures of territorialization aiming to control people and resources are, for example, administrative reordering or enforced settlements of mobile social groups in rural areas. State authority and domination are secured through territorial control, whereby local actors might accept or ignore state practices of territorial control, or fight against them (Berry 2009: 24). Territory therefore cannot be reduced to a fixed and static resource, which is contested in distributional conflicts (see Featherstone 2004: 703), but it is, at the same time, a material reference for authority as well as for the construction of collective identity. Territorialization as an analytical concept enables us to capture these multiple and nested functions of territory by investigating the relations between space and conflict rather than assuming that one influences the other.

Scale

Drawing on the work of Smith (1984) and Taylor (1982), among others, radical geographers challenge the tendency to take the nation-state for granted as a central unit of analysis on which analytical scales are defined as being located either below it (subnational) or above it (international). Their aim is to overcome the hierarchically connoted dualism that locates causal factors on the global and social agency on the local scale. They thus question equating 'the local' with the everyday and locality, and 'the global' as an abstract space in opposition to it (Escobar 2001: 155). From this perspective, neither power nor conflict are located in definite, bounded spaces and it is no longer possible to imagine scales (global, national, local) as hierarchical and discrete bounded entities. For conflict studies, scale is a useful analytical concept because it allows for a focus on the social construction of conflicts, as well as on how they are produced by, and productive of, various agents (Engels 2015).

Towers provides a helpful distinction for analysing these processes of scale construction: he differentiates conceptually between 'scales of regulation' and 'scales of meaning' (Towers 2000: 26). While the former is the institutional scale of political regulation (municipality, nation-state, European Union (EU), UN, etc.), the latter refers to a product of social interpretation. Both concepts focus on how scales are socially constructed, negotiated and changed, thus emphasizing the actors' scalar practices rather than 'scale itself' (Neumann 2009: 399). As has been pointed out in the controversy over Marston et al.'s (2005) 'human geography without scale', the question is generally not 'whether scale "exists"' (Hoefle 2006: 238) but what it means. If we put the scales of regulation at the centre of our analysis we investigate struggles over the scalar arrangements themselves, and how actors try to shift political power and competence to different scales. Conflicts over land and territory are also conflicts over the question where (at what scale) and by whom power is exercised. Focusing on the scales of meaning asks, for instance, how conflict actors discursively locate a particular social problem, its causes and possible solutions. This itself is part of a political conflict. As Smith put it, 'The scale of struggle and the struggle over scale are two sides of the same coin' (1992: 74).

Conclusion

By way of conclusion we ask what it means to study conflicts from a spatial perspective. Our central argument is that practices of discursive and material spatialization provide a rich field of investigation. Focusing on the production of space in Lefebvre's sense enables us to investigate how the material and the social are related to each other and thus to denaturalize scale, place and other spatial entities (see The Space of Democracy and the Democracy of Space Network 2009: 580). This is not to suggest that material influences do not matter or to deny their principal autonomy, but to call into question their conceptualization as external factors that influence or even determine social (conflict) action. Our intention is not to develop a new variant of determinism based on the assumption of space as a social construction. Rather, it is our contention that the analytical focus should be on the interplay of conflict action on the one hand and the social and political production of space on the other.

With regard to spatial analysis in the study of contentious politics, Martin and Miller suggest:

> Should space be thought of as a variable – distance, for instance – to be added on to an otherwise aspatial analysis? Should space be thought of in terms of place-specific forms of identity, e.g., neighbourhood identity or nationalism, separate from 'non-spatial' forms of identity such as gender, race, and class? ... There is precedent for the adoption of each of these conceptualizations, and in certain instances each may yield important insights.
>
> (2003: 144)

This also holds true for spatial analysis in conflict studies. If we assume that space and social action are inseparably interwoven, it is hardly possible to 'add space and stir' to an 'otherwise aspatial analysis' of violent conflict. However, this is what most existing studies seem to do: to add physical and material features and variables to an analysis of conflict, which is based on assumptions of mostly linear causal relations. As argued in this chapter, if studies do not theoretically reflect on the relationship between space and conflict, they risk falling back on deterministic assumptions regarding the material conditions for social action, and may have difficulties conceptualizing how physical and material features are mediated socially and politically. Yet instead of simply inverting this approach by focusing solely on social and cultural dynamics, we suggest that conflict as social action has a mutually constitutive relationship with space. From this perspective, we suggest analysing spatial conditions that are produced and reproduced by social conflicts and, at the same time, enable and limit further action. Space does not determine conflict, but conflict produces, structures and restructures space. Building upon this assumption, we can thus ask what kind of social action enables or hinders certain spatial conditions.

From the perspective of conflict studies, therefore, conflict as a social practice contributes to the (shifting) meanings of space and symbolic, identity-related factors at different levels. The analysis of the relationship between space and conflict has to identify the interaction between physical materiality, its social and cultural ascription, and conflict dynamics. Social conflict therefore has constructive power for the physical and material as well as the sociopolitical dimension of space.

Notes

1. We understand 'identity' as a category of the socially and culturally constructed perception and 'belonging' that manifests itself in social practices and collective action. It is an inherently relational category, meaning that the construction of the 'self' necessarily refers to a constructed 'other' (for a detailed and critical discussion, see Brubaker and Cooper 2000).
2. Whereas structural data give information about economic, political, demographic or military development, vector data represent geographical information based on points, lines, curves and shapes or polygons (e.g. cities, transport routes, resource occurrence, surface areas and settlement or population density).
3. Without doubt, the availability of relevant and reliable data is a challenge, particularly on the substate level. This holds true for quantitative and qualitative studies alike. Nevertheless, from a methodological perspective it is problematical to choose variables primarily on the basis of data availability.
4. For instance, this is expressed in the famous quote that 'the class struggle is inscribed in space' (Lefebvre 1994: 5).

References

Adger WN (1999) Social Vulnerability to Climate Change and Extremes in Coastal Vietnam. *World Development* 27(2): 249–269.

Adger WN (2006) Vulnerability. *Global Environmental Change* 16(3): 268–281.

Agnew J (1994) The Territorial Trap: The Geographical Assumptions of International Relations Theory. *Review of International Political Economy* 1(1): 53–80.

Baechler G, Spillmann KR and Suliman M (eds) (2002) *Transformation of Resource Conflicts: Approach and Instruments*. Bern: Peter Lang.

Benjaminsen T and Boubacar B (2009) Farmer-herder Conflicts, Pastoral Marginalisation and Corruption: A Case Study from the Inland Niger Delta of Mali. *The Geographical Journal* 175(1): 71–81.

Benjaminsen TA, Alinon K, Buhaug H and Buseth JT (2012) Does Climate Change Drive Land-use Conflicts in the Sahel? *Journal of Peace Research* 49(1): 97–111.

Berry S (2009) Property, Authority and Citizenship: Land Claims, Politics and the Dynamics of Social Division in West Africa. *Development and Change* 40(1): 23–45.

Blimes RJ (2006) The Indirect Effect of Ethnic Heterogeneity on the Likelihood of Civil War Onset. *Journal of Conflict Resolution* 50(4): 536–547.

Brenner N (2008) Tausend Blätter: Bemerkungen zu den Geographien ungleicher räumlicher Entwicklung. In: Wissen M, Röttger B and Heeg S (eds) *Politics of Scale. Räume der Globalisierung und Perspektiven emanzipatorischer Politik*. Münster: Westfälisches Dampfboot, pp. 57–84.

Brubaker R and Cooper F (2000) Beyond 'Identity'. *Theory and Society* 29(1): 1–47.

Buhaug H (2010) Dude, Where's My Conflict? LSG, Relative Strength, and the Location of Civil War. *Conflict Management and Peace Science* 27(2): 107–128.

Buhaug H and Gates S (2002) The Geography of Civil War. *Journal of Peace Research* 39(4): 417–433.

Buhaug H, Gates S and Lujala P (2009) Geography, Rebel Capability, and the Duration of Civil Conflict. *Journal of Conflict Resolution* 53(4): 544–569.

Buhaug H and Rød JK (2006) Local Determinants of African Civil Wars, 1970–2001. *Political Geography* 25(3): 315–335.

Deligiannis T (2012) The Evolution of Environment-Conflict Research: Toward a Livelihood Framework. *Global Environmental Politics* 12(1): 78–100.

Diehl PF (1991) Geography and War. A Review and Assessment of the Empirical Literature. *International Interactions* 17(1): 11–27.

Dietz K, Engels B and Pye O (2015) Territory, Scale and Networks: The Spatial Dynamics of Agrofuels. In: Dietz K, Engels B, Pye O and Brunnengräber A (eds) *The Political Ecology of Agrofuels*. Abingdon: Routledge, pp. 34–52.

Duncan JS, Johnson NC and Schein RH (eds) (2004) *A Companion to Cultural Geography*. Malden: Blackwell.

Elden S (2007) There Is a Politics of Space Because Space Is Political: Henri Lefebvre and the Production of Space. *Radical Philosophy Review* 10(2): 101–116.

Engels B (2015) Contentious Politics of Scale: The Global Food Price Crisis and Local Protest in Burkina Faso. *Social Movement Studies* 14(2): 180–194.

Escobar A (2001) Culture Sits in Places: Reflections on Globalism and Subaltern Strategies of Localization. *Political Geography* 20(2): 139–174.

Fearon JD (2005) Primary Commodity Exports and Civil War. *Journal of Conflict Resolution* 49(4): 483–507.

Featherstone D (2004) Spatial Relations and the Materialities of Political Conflict: the Construction of Entangled Political Identities in the London and Newcastle Port Strikes of 1768. *Geoforum* 35(6): 701–711.

Fox J (2004) The Rise of Religious Nationalism and Conflict: Ethnic Conflict and Revolutionary Wars, 1945–2001. *Journal of Peace Research* 41(6): 715–731.

Gartzke E and Gleditsch KS (2006) Identity and Conflict: Ties That Bind and Differences That Divide. *European Journal of International Relations* 12(1): 53–87.

Gates S (2002) Recruitment and Allegiance: The Microfoundation of Rebellion. *Journal of Conflict Resolution* 46(1): 111–130.

Gleditsch NP (1998) Armed Conflict and the Environment: A Critique of the Literature. *Journal of Peace Research* 35(3): 381–400.

Gleditsch NP (ed.) (1997) *Conflict and the Environment*. Dordrecht: Kluwer.

Hauge W and Ellingsen T (1998) Beyond Environmental Scarcity: Causal Pathways to Conflict. *Journal of Peace Research* 35(3): 299–317.

Hendrix CS and Glaser SM (2007) Trends and Triggers: Climate, Climate Change and Civil Conflict in Sub-Saharan Africa. *Political Geography* 26(6): 695–715.

Hoefle SW (2006) Eliminating Scale and Killing the Goose That Laid the Golden Egg? *Transactions of the Institute of British Geographers* 31(2): 238–243.

Homer-Dixon TF (1999) *Environment, Scarcity, and Violence*. Princeton: Princeton UP.

IPCC (2007) *Climate Change 2007*. IPCC fourth assessment report. Geneva: IPCC. Available at: http://www.ipcc.ch/publications_and_data/publications _ipcc_fourth_assessment_report_synthesis_report.htm (accessed 25 June 2015).

Jessop B, Brenner N and Jones M (2008) Theorizing Sociospatial Relations. *Environment and Planning D: Society and Space* 26(3): 389–401.

Kalyvas SN (2006) *The Logic of Violence in Civil War*. Cambridge: Cambridge University Press.

Kalyvas SN (2012) Micro-Level Studies of Violence in Civil War: Refining and Extending the Control-Collaboration Model. *Terrorism and Political Violence* 24(4): 658–668.

Le Billon P (2001) The Political Ecology of War: Natural Resources and Armed Conflicts. *Political Geography* 20(5): 561–584.

Lefebvre H (1994) *The Production of Space*. Oxford: Blackwell.

Lujala P, Gleditsch NP and Gilmore E (2005) A Diamond Curse? Civil War and a Lootable Resource. *Journal of Conflict Resolution* 49(4): 538–562.

Lund C (2011) *Land Rights and Citizenship in Africa*. NAI Discussion Paper 65. Uppsala: Nordiska Afrikainstitutet.

Marston SA, Jones III JP and Woodward K (2005) Human Geography without Scale. *Transactions of the Institute of British Geographers* 30(4): 416–432.

Martin DG and Miller B (2003) Space and Contentious Politics. *Mobilization: An International Journal* 8(2): 143–156.

Massey DB (1984) Introduction: Geography Matters. In: Massey DB and Allen J (eds) *Geography Matters! A Reader*. Cambridge: Cambridge University Press, pp. 1–11.

Mayer M (2008) To What End Do We Theorize Sociospatial Relations? *Environment and Planning D: Society and Space* 26(3): 414–419.

Most BA and Starr H (1980) Diffusion, Reinforcement, Geopolitics, and the Spread of War. *The American Political Science Review* 74(4): 932–946.

Neumann RP (2009) Political Ecology: Theorizing Scale. *Progress in Human Geography* 33(3): 398–406.

Newman D (2010) Territory, Compartments and Borders: Avoiding the Trap of the Territorial Trap. *Geopolitics* 15(4): 773–778.

O'Loughlin J, Holland EC and Witmer FDW (2011) The Changing Geography of Violence in Russia's North Caucasus, 1999–2011: Regional Trends and Local Dynamics in Dagestan, Ingushetia, and Kabardino-Balkaria. *Eurasian Geography and Economics* 52(5): 596–630.

Raleigh C and Kniveton D (2012) Come Rain or Shine: An Analysis of Conflict and Climate Variability in East Africa. *Journal of Peace Research* 49(1): 51–64.

Raleigh C and Urdal H (2007) Climate Change, Environmental Degradation and Armed Conflict. *Political Geography* 26(6): 674–694.

Raleigh C, Witmer F and O'Loughlin J (2010) A Review and Assessment of Spatial Analysis and Conflict: The Geography of War. In: Denemark RA (ed.) *The International Studies Encyclopedia Vol X*. Oxford: Wiley-Blackwell, pp. 6534–6553.

Reuveny R (2007) Climate Change-induced Migration and Violent Conflict. *Political Geography* 26(6): 656–673.

Ross ML (2004) What Do We Know about Natural Resources and Civil War? *Journal of Peace Research* 41(3): 337–356.

Sack RD (1986) *Human Territoriality. Its Theory and History*. Cambridge: Cambridge University Press.

Salehyan I (2008) From Climate Change to Conflict? No Consensus Yet. *Journal of Peace Research* 45(3): 315–326.

Siverson RM and Starr H (1991) *The Diffusion of War. A Study of Opportunity and Willingness*. Ann Arbor: University of Michigan Press.

Smith N (1984) *Uneven Development: Nature, Capital and the Production of Space*. Oxford: Blackwell.

Smith N (1992) Geography, Difference and Politics of Scale. In: Doherty J, Graham E and Malek M (eds) *Postmodernism and Social Science*. London: Palgrave Macmillan, pp. 57–79.

Soja EW (1989) *Postmodern Geographies. The Reassertion of Space in Critical Social Theory*. New York: Verso.

Soja EW (1996) *Thirdspace: Journeys to Los Angeles and Other Real-and-imagined Places*. Cambridge: Blackwell.

Starr H (1991) Joining Political and Geographic Perspectives: Geopolitics and International Relations. *International Interactions* 17(1): 1–9.

Stephenne N, Burnley C and Ehrlich D (2009) Analyzing Spatial Drivers in Quantitative Conflict Studies: The Potential and Challenges of Geographic Information Systems. *International Studies Review* 11(3): 502–522.

Taylor PJ (1982) A Materialist Framework for Political Geography. *Transactions of the Institute of British Geographers* 7(1): 15–34.

Taylor PJ (1996) Embedded Statism and the Social Sciences: Opening Up to New Space. *Environment and Planning A* 28(11): 1917–1928.

The Space of Democracy and the Democracy of Space Network (2009) What Are the Consequences of the 'Spatial Turn' for How We Understand Politics Today? A Proposed Research Agenda. *Progress in Human Geography* 33(5): 579–586.

Theisen OM (2012) Climate Clashes? Weather Variability, Land Pressure, and Organized Violence in Kenya, 1989–2004. *Journal of Peace Research* 49(1): 81–96.

Towers G (2000) Applying the Political Geography of Scale: Grassroots Strategies and Environmental Justice. *The Professional Geographer* 52(1): 23–36.

Vandergeest P and Peluso NL (1995) Territorialization and State Power in Thailand. *Theory and Society* 24(3): 385–426.

Vasquez JA (2009) *The War Puzzle Revisited*. Cambridge: Cambridge University Press.

Warf B and Arias S (eds) (2009) *The Spatial Turn: Interdisciplinary Perspectives*. Abingdon: Routledge.

Wimmer A and Glick Schiller N (2002) Methodological Nationalism and Beyond: Nation-state Building, Migration and the Social Sciences. *Global Networks* 2(4): 301–334.

Wisner B, Blaikie P, Cannon T and Davis I (1994) *At Risk: Natural Hazards, People's Vulnerability, and Disasters*. Abingdon: Routledge.

Wucherpfennig J, Weidmann NB, Girardin L, Cederman LE and Wimmer A (2011) Politically Relevant Ethnic Groups Across Space and Time: Introducing the GeoEPR Dataset. *Conflict Management and Peace Science* 28(5): 423–437.

2
Territoriality in Civil War: The Ignored Territorial Dimensions of Violent Conflict in North Kivu, DRC

Martin Doevenspeck

Introduction

On the morning of 28 October 2008, the eastern Congolese rebel movement, the CNDP, occupied the both strategically and symbolically important district capital of Rutshuru in the eastern Congolese province of North Kivu. The rebels drove out the government troops, continued their advance on the city of Goma on the shores of Lake Kivu, and thus helped to swell the numbers of refugees – already several hundred thousands. At the same time, at a distance of about 20 km from Rutshuru as the crow flies, in the area controlled by the rebels and within hearing distance of the shooting, some 4,000 people from the surrounding villages were busy in the market of Bunagana on the Congolese-Ugandan border. Nobody seemed bothered by the noise of the fighting or let it stop them from carrying on their business. Feeling rather nervous, I asked why everyone was so relaxed, and a woman selling secondhand shoes shrugged her shoulders and said:

> There's no problem, the war is far away. The FARDC (the Congolese army [Forces Armées de la République Démocratique du Congo/Armed Forces of the Democratic Republic of Congo]) will run away, as usual, and leave their weapons for the CNDP. The rebels are in control here, but we are safe. And that is the most important thing.[1]

Four-and-a-half years later, in March 2013, troops of the M23 rebels fled from Rutshuru after an internal split and fighting between two rival fractions, leaving the 100,000 inhabitants to fend for themselves. At practically the same time, various militias entered the town, which the M23, the successor of the CNDP, had recaptured just eight months earlier, together with the former eastern core area of the CNDP on the Congolese-Rwandan-Ugandan border. The administrative order established during this time and the security model of the M23 collapsed within a few hours of the withdrawal of the troops. 'Here people in uniform are running around everywhere and robbing us. The M23 has gone away without warning us, they've let us down.'[2]

How can it be explained that while soldiers are dying and civilians are fleeing, people just a few kilometres away feel secure enough to carry on their normal business and gossip with each other in the market? And what is the connection between the ability and the will to exercise direct territorial control and the success of a politicomilitary rebellion?

Starting from these questions, and embedded in a theoretical outline of territory, this chapter aims to analyse from a political geography perspective both successful and unsuccessful modes of the production of territoriality by a rebel movement. This production of territoriality is one of several territorial parameters that may shape the outcome of violent conflict and thus an important, yet often ignored, factor in peace and conflict studies. This applies especially when it comes to understanding the efficacious establishment of political order by non-state violent actors in civil war. The geography of violence outlined in this chapter thus centres on the phenomenon of an imagined enclave of fragile local peace and predictable political order in civil war. In this approach, space is considered not only as an arena for conflict but, via the concepts of territory and territoriality, also as an analytical approach to it.

To illustrate my argument I draw mainly on empirical findings from research conducted in October and November 2008 and in September 2012 in the rebel-held Rutshuru district. A third field trip to the rebel territory in March 2013 had to be cancelled at short notice when fighting broke out between the two rival fractions of the M23. However, I tried to follow events by conducting telephone interviews from Goma (DRC) and Gisenyi (Rwanda). In 2008, after a short period of suspicion following my arrival in Bunagana, I was allowed to travel freely in the area. I spoke to rebels of the political and military wings of both the CNDP and the M23, to ordinary soldiers, policemen, civil servants, peasants, teachers and market women. In 2008 I also had lengthy discussions with the commander of the Indian mobile base of MONUC (the UN Organization Mission in the DRC; now MONUSCO, Mission de l'Organisation des Nations Unies pour la stabilisation en République démocratique du Congo/the UN Organization Stabilization Mission in the DRC), the peacekeeping mission of the UN in the DRC, then based near Chengerero. I attended local trials, markets and military parades, inaugurations of mayors selected by the rebel administration and other symbolic events where territoriality was produced.

After showing how territorial dimensions are ignored in attempts to understand and deal with violent conflict in North Kivu, and mapping the concepts of territory and territoriality, the context is introduced with a special focus on the CNDP and the M23, the politicomilitary movements of the Congolese Tutsi. The main part of the chapter then traces the institutionalization of a territory and the production of territoriality by these rebels. By concentrating on borders, symbols and institutions, it covers the patterns, results and (hidden) costs of the production of territoriality. The chapter concludes by bringing together theoretical reflections on territory and territoriality and empirical findings, in order to present a different understanding of this part of violent conflict in North Kivu.

Ignorance of the territorial dimension of warfare in eastern DRC

Besides political science approaches to civil war in the DRC on a macrolevel, which mostly do not even touch upon the issue of armed groups, the dominant representation of such a group in eastern DRC is either that of a brutal, pillaging and apolitical militia, or that of a compliant proxy of a neighbouring state that pursues its political and economic interests by supporting rebel groups. As Radil and Flint point out in their study of war diffusion in the DRC, little is known about the internal structure, political agendas and territorial practices of violent non-state actors in this conflict: 'While there is a growing recognition of the need to shift the unit of analysis beyond the state for studies of conflict, little systematic or longitudinal data is available about the characteristics of these groups themselves' (Radil and Flint 2013: 192).

This also applies to the rebellions by groups of Congolese Tutsi in North Kivu who have presented the most serious challenge, both militarily and politically, to the government of President Joseph Kabila since his coming into power ten years ago. It is well known that both the CNDP and the M23 were supported logistically, and on certain occasions also militarily, by Rwanda and to a lesser extent Uganda, both of which had their own particular agendas. However, an isolated view of the production of domestic sovereignty by means of various extraterritorial actions (support of rebel groups, undercover proactive counterinsurgency, etc.) by those neighbouring states, which undoubtedly bear their specific responsibility for the human tragedy going on in the region, falls short in understanding this particular segment of violent conflict in North Kivu. Leaving aside the fact that there is hardly any armed group in eastern DRC that does not benefit from external support in some form or other, there seems to be a general inclination towards externalizing the causes of war or reducing them to the 'resource war' argument. This tendency might be considered a form of ignorance, which can be exploited as a resource by those who have an interest in obscuring important aspects of violent conflict in this region, or by those who can live more comfortably without recognizing the complex combination of conflict causes and drivers.

Much has been said, for example, about the shortcomings of interpreting violent conflict in the DRC as a resource war. That raw material extraction had the effect of prolonging the conflict during the global coltan boom at the turn of the millennium, the time of the so-called Second Congo War, is undisputed. And in recent years, campaigns by NGOs have repeatedly underlined the importance of raw materials for financing the wars and as a motivation for violence in eastern DRC, and have led to a range of measures to restrict trade involving so-called 'conflict minerals'.

Such measures are based in part on the assumption that prohibiting trade with conflict minerals would also put an end to the violence. It is true that, like most other conflicts, this one also has an important and highly complex economic dimension, but from an analytical point of view the widespread focus on raw materials as the cause is too shortsighted. A number of authors have pointed out that this conflict has a variety of causes and drivers (see Doevenspeck 2011; Prunier 2009; Vlassenroot and Raeymaekers 2009) and have argued that it cannot simply be interpreted as a fight for resources. The following factors have been identified to explain the armed conflicts which have been going on with changing coalitions and varying degrees of intensity for almost 20 years: ethnicized struggles over access to land and to political institutions, questions of citizenship rights and national identity, the link to the Rwandan conflict dynamic and the marginalization of this border region by the central state in the context of general government failure.

However, the picture of the DRC as a victim of its own resources is tenacious and is perpetuated both by the government and by the media. In this regard, the Congolese resource war narrative would be a striking example for the study of what McGoey (2012) has called 'strategic unknowns'. She and others (e.g. Gross 2010; High et al. 2012) in the emerging field of the study of ignorance point to the need for studying the general advantages, and concrete and material profits, of deflecting, denying, covering and obscuring knowledge. In this view of intentional non-knowledge as a productive force that is able to mobilize both material and immaterial resources, the keeping up of the resource war interpretation maintains a selective understanding of violent conflict and allows the denial of unsettling and uncomfortable facts. These include an unwillingness on national and regional levels to develop functional state structures, the conflicting role of power brokers on different scales, and the inhomogeneity of the so-called 'local communities' and their recalcitrance towards global models of conflict transformation and their creativity in translating those models (Autesserre 2010, 2014).

The lack of territorial control by the state combined with the erosion of sovereignty caused by international intervention has increased a security vacuum, which is filled by a multitude of armed groups that compete on profitable security markets. In this chapter I will concentrate on the peculiar impact of this uncomfortable fact: the establishment and maintenance of territory, or the production of territoriality, by a rebel group. I begin by briefly introducing the theoretical concepts of territory and territoriality.

Territory and territoriality

If we consider the simultaneity of de- and reterritorialization processes – for example, the deterritorialized US drone war or the creation of a new territory by the so-called 'Islamic State' (IS) that has razed colonial borders, and the linked rescaling and restructuring of state power – it is obvious that territory matters when it comes to understanding war. The fact that violence seldom prevails at all times and in all parts of an area affected by civil war (Korf et al. 2010), and that security can be established by non-state violent actors in the areas controlled by them (Schlichte 2009), also confirms the importance of territorial formations for explaining the spatiotemporal structural dynamics of war and peace.

Even if it is widely acknowledged that territory means a segment of bounded space at where a claim of sovereignty is targeted, it is, above all, an ambiguous term – ambiguous in the sense that despite the danger of a 'territorial trap' (Agnew 1994) and post-structuralist objections, there are not only visible and material but also hidden and fluid borders, spatial incoherence, fragmented sovereignty, multiple nested identities and, not least, various encapsulated scales and materializations beside the state. All of these need to be considered in order to understand the culturally and socially constructed nature of territories and the power structures inherent to them. However, now that political geography has overcome its state-centred paradigm with regard to the political dimensions of space and the spatial dimension of politics, it thinks of territory not as pre-existing but as something that becomes (Knight 1982). Territory, as a portion of bounded space with material (land), functional (control) and symbolic (identity) dimensions, depends for its production and perpetuation on a multitude of technologies, practices and discourses. In this sense it can be understood as a social process. The process of forming and establishing territories, as well as their identification through social agency and awareness, can be considered the institutionalization of territories. This institutionalization can be understood with the help of three abstractions representing different aspects of territory formation (Paasi 1995, 2003). First is the territorial shape, which means the construction of physical or symbolic boundaries as the basic element of the practice of territoriality. Second is the symbolic shape, which includes symbols such as flags or emblems, and social practices in which these elements come together, such as military parades or manifestations. Last is the institutional shape, meaning institutionalized practices which reproduce, for example, boundaries and symbols and their relevance: administration, politics, economy, culture or the education system. The institutionalization of territories means that territoriality is produced with an understanding as a strategy to control people and things, and the relationship between them, through the control of an area. Territoriality is therefore the most basic spatial form of power (Sack 1986).

Thus territory, its production and perpetuation, and the resulting consequences and effects, cannot be taken for granted, and the sociotechnical practices associated with the three abovementioned abstractions must be understood as being interlinked. That these practices can succeed just as much as fail has implications for the understanding of territories, which may be porous, uneven, changeable or historically contingent, and, of course, may also disappear (Painter 2010). A territory is permanent work, never complete but always in the process of creation. This work is liable to tensions and contradictions, and it can go wrong.

Despite justified criticism of an overly narrow concept of territory (Elden 2009, 2010), and despite an increasing deterritorialization of state and non-state power (Agnew 2005), in the case of the rebellions studied here, we have to consider the supposedly 'old', territorially based concept of power if we are to understand the constitution of an enclave of local peace. In this sense, the empirical example presented here reveals the importance of non-Western contexts and does not conform to the relatively restricted field of contexts to which the theoretical assumptions in conflict research, political geography and political science, which often claim universal applicability, are related (Sidaway 2008).

The CNDP and the M23 in the East Congo civil war

Political and military constellations change quickly in the DRC, especially in the two Kivu provinces. Changing and unexpected alliances, and the proliferation of violent actors in the context of negotiating geopolitical strategies and economic interests with neighbouring countries, as well as old and new hegemonial powers, do not make it easy to understand the politicomilitary situation at any particular time (Vlassenroot and Raeymaekers 2009). Within the confusing array of armed groups in the Kivu provinces, the politicomilitary Congolese Tutsi organizations, the CNDP and the M23, stand out because of their superior strength. This was demonstrated by the M23 when it captured the provincial capital of Goma in November 2012. The CNDP also inflicted several heavy defeats on the FARDC between 2006 and 2009.

Yet today neither the CNDP nor the M23 exists in the form described here. However, despite the transience of political structures, they are only seldom completely replaced by new ones. Rather, they show a distinct persistence and form a new layer in the complex sediment of political orders in eastern DRC. Therefore it is worth taking a look at the history and the political aims of the CNDP and the M23 in order to provide an outline of the perpetuation of the violent conflict in North Kivu.

Violent conflict around the triangle of identity, local power and access to land in North Kivu began in the early 1990s, culminating in the heavy fighting of 1993. In the course of the civil war and genocide in Rwanda, the conflicts took on a new dimension when 2 million Rwandan Hutu fled into the Kivu provinces. Among the refugees were many thousands of armed Rwandan soldiers and militiamen, who reorganized in the refugee camps to attack the new regime in Rwanda. As a result, Congolese Tutsi, fearing this armed group, fled to Rwanda, as well as to Uganda and Tanzania. Partly in order to combat the Rwandan Hutu militias, the new Government of Rwanda led the war from 1996 to 1997 in which the longstanding dictator, Mobutu, was overthrown.

Many Congolese Tutsi who had fled, fought on the side of the Rwandan forces and were later to form the core of the military leadership of the CNDP. Between 1998 and 2003, Rwanda, via its proxy, the RCD-Goma (Rassemblement Congolais pour la Démocratie/Congolese Rally for Democracy), controlled large parts of eastern DRC. The RCD-Goma consisted primarily of Congolese Hutu and Tutsi, and it can be regarded as a precursor to the CNDP. After the official signing of the Sun City peace agreement in South Africa in 2003, the RCD-Goma became a political party, participated in the transitional government of President Joseph Kabila and sank into insignificance after the 2006 elections. The resulting lack of political representation of the Kinyarwanda-speaking people of North Kivu left a vacuum, which was cleverly exploited by the CNDP. Shortly before the elections, it had been founded by Laurent Nkunda who, together with other central figures in the CNDP, had participated in all wars in the region since the Rwandan civil war, and now wanted to assert the interests of the Kinyarwanda-speaking Congolese, especially the Tutsi, if necessary by violent means (Scott 2008). The beginning of violent clashes was justified chiefly by the threat posed to the Congolese Tutsi by the FDLR (Forces Démocratiques de Libération du Rwanda/Democratic Forces for the Liberation of Rwanda), the restructured Rwandan Hutu militia in the DRC.

The intention of the CNDP was to give military backing to its demands for the return of Tutsi refugees and for their political representation. The CNDP received logistic, and sometimes also military, support from Rwanda. As former CNDP and M23 members, bitter and disillusioned, now acknowledged, those in power in Rwanda were not really interested in the situation of the Congolese Tutsi but in ensuring their support in safeguarding Rwanda's complex economic and security interests in eastern DRC.[3]

Although the CNDP raised claims on the national level and was able to partially integrate an ethnically mixed opposition to Kabila's government, its central political goals were ethnically motivated (Stearns 2008). This is exemplified by the discourse on autochthony. The following quotation is from an interview with Rene Abandi, a lawyer and one of the intellectual masterminds of the CNDP and, to a lesser extent, of the M23. He fled from Masisi in the early 1990s, worked in Italy, joined the CNDP in 2005 and at the time of the interview was something akin to its foreign minister.

Take Goma for example. Up to the 1970s or 1980s, everyone here spoke Kinyarwanda. That was before all these people came from

Bukavu, Butembo or Kisangani, who now call us foreigners. During the peace conference in Goma I said to them: Look, when I was in school, I saw you arriving. Even my little brother saw you arriving here. And now you say that the farm of my father and my grandfather doesn't belong to me? If this doesn't stop, you will have no peace here in the east.[4]

For a long time the CNDP was the strongest military power in North Kivu and fought not only against the army but also against the FDLR, the anti-Rwandan Mayi-Mayi and other militias. Up to 2009 the CNDP ruled large areas in North Kivu, developed a comparatively well-functioning administration and established an unrestricted monopoly on violence with no clashes within the rebel territory.

When CNDP troops threatened to take Goma in November 2008, international pressure on Rwanda led to a political rapprochement between Joseph Kabila and the Rwandan president, Paul Kagame. In a spectacular manoeuvre, his former supporters in Rwanda arrested Laurent Nkunda in January 2009 with the help of his internal rival, Bosco Ntaganda. This led to a split of the CNDP into a faction loyal to Nkunda, led by Sultani Makenga, and another loyal to Ntaganda. Ntaganda, against whom a warrant of arrest had been issued by the International Criminal Court in The Hague, signed a peace agreement between the CNDP and the Congolese Government on 23 March 2009. The deal allowed CNDP combatants to be integrated into the FARDC and promised that Tutsi refugees in Rwanda and Tanzania would be able to return to the DRC, and that there would be politicoadministrative decentralization and local elections.

Ntaganda was given the rank of general in the FARDC and led almost all military operations in the provinces of North and South Kivu up to the end of 2011. He held important positions in the FARDC with former CNDP officers and stationed loyal troops in strategically important positions. During the presidential elections in 2011, Ntaganda supported President Kabila with the aim of getting former CNDP members into public office. UN reports accused Ntaganda of ballot-rigging and using troops to intimidate voters. Donor countries used the accusation of electoral fraud to urge Kabila to carry out reforms and demanded that Ntaganda should be arrested and turned over to the International Criminal Court. This triggered the desertion of CNDP soldiers in spring 2012. After initial defeats at the hands of the government army, the troops loyal to Ntaganda and those of Sultani Makenga came together at the Congolese-Rwandan-Ugandan border. Supported by Rwanda and at least tolerated by Uganda, they relatively quickly conquered the former eastern core area of the CNDP between Bunagana and Rutshuru, and even temporarily occupied the city of Goma. But finally, in November 2013, they were defeated by the new UN Force Intervention Brigade. The name of this new rebellion, the M23, was a reference to 23 March 2009, the date of the peace agreement, which the rebels complained had not been complied with.

Because of the unbridgeable gap between the Makenga faction and the group loyal to Ntaganda, the M23 was split from the beginning. Although the rivalry was kept at bay for a while, it exploded violently in March 2013. The two groups fought each other for days, ending with the defeat of Ntaganda, who fled into the US Embassy in Kigali. A huge loss of political legitimacy was added to the military weakness the fighting had caused. Towns such as Rutshuru were at times abandoned and fell into the hands of enemy militias. This clearly revealed that the military power of the M23 was inferior to that of the CNDP, and also showed that it was not able to hold a monopoly on violence, which would be necessary to legitimate the alternative state it had built up on the model of the CNDP. The rebellion was badly prepared, both militarily and, since it lacked a proper political programme and alliances worth mentioning with the political opposition in the country, also politically. I will now take a look at the practices of the production of territoriality by the CNDP and the M23.

Borders, symbols and institutions: The production of rebel territoriality

Territoriality is produced not just by the state but also by many different actors, in this case by rebels. Territory as a social process consists of material elements such as land, functional elements such as control over a space, and symbolic dimensions such as identity. The process of the creation, establishment and maintenance of territories and their identification through social practice can be referred to as the institutionalization of territories, or the production of territoriality. Territoriality as political technology, as a strategy employed by individuals or groups to gain control over people, things and the relations between them by means of control over a territory, is thus a fundamental spatial form of power (Sack 1986).

The process of institutionalizing a territory – in other words, the production of territoriality – will be described in terms of the three abstractions presented earlier, which illustrate different aspects of the formation of territories: territorial, symbolic and institutional shape.

Territorial shape

Territorial shape refers to the construction of physical and symbolic borders as a basic element of the practice of territoriality. In the case of the CNDP, besides the borders constituted at any time by the front lines, this was manifested especially in the management of the border with Uganda at Bunagana. In 2008, anyone who entered the DRC here without being aware of the situation – as, for instance, a group of American tourists on their way to see the mountain gorillas in Virunga National Park – did not necessarily notice immediately that they were now in a rebel-controlled territory. Here, just as at other border crossing points in the region, uniformed officials recorded the name, occupation and reason for visit of every traveller and asked whether they had anything to declare. Only a closer look at the flag – a hand holding a flaming torch on a light blue background – revealed a first symbol of the territoriality exercised here by the rebels. The border at Bunagana was important not only because a large part of the consumer goods shipped from Dubai or China to Mombasa entered the country here; the rebels were also able to use the crossing point as a demonstration of the fact that they were able to organize the border according to their ideas of efficiency and transparency, in contrast with the one in Goma (Doevenspeck 2011):

> In Goma you will be fleeced. It's worst at the Petite Barrière [one of two official crossing points between Goma, DRC and the Rwandan town of Gisenyi]. Everyone wants to have something from you … They aren't interested in national security, only in money.[5]

In addition to the border with Uganda, the front line constituted a mobile border between the CNDP and M23 territory and the government-controlled areas. The topography of the border region around the Virunga volcanoes enabled the rebels to better control the transport infrastructure around their territory from higher ground. And during a patrol in the hills east of the important Goma–Rutshuru road, it could be seen that at intervals of just a few hundred metres, small mobile groups of soldiers in well-camouflaged shelters were keeping watch over almost the entire length of the border to prevent attacks by FDLR combatants. The Indian commander of the mobile base of the UN peace troops (MONUC/MONUSCO) in the rebel area, a fenced-in and heavily guarded counterenclave near Chengerero, which was tolerated by the rebels in view of their international image, commented:

> There is no fighting here, there is no shooting here. The only thing we do is going out on daily patrols, but everything is quiet ... I tell Castro[6] when we set off and he calls me if they want anything from us.'[7]

Symbolic shape

This covers symbols such as flags, emblems, statues and social practices, such as parades and assemblies in which these elements come together. In addition to flags and emblems at the border and on administrative buildings, the symbolic aspect of the territoriality of the rebellion included weapon displays. Thus, after the capture of Rumangabo by the rebels on 8 October 2008, the weapons seized from the FARDC were displayed to the people in Chengerero and to the Indian UN soldiers who were also present. Another format was ideological training on the policies and goals of the CNDP and later of the M23, which was compulsory for the local people. Topics included the historical roots of the civil war, a fundamental critique of the Kabila government, and the rebels' conception of good government. During the time of the CNDP, this training was accompanied by the introduction of local reconciliation committees. If the Congolese Tutsi were subjected to expulsion and violence in the Rutshuru region up to the invasion by Rwanda in 1996, it was subsequently the majority Congolese Hutu who were persecuted and killed, since the Rwandan army often equated them with the Rwandan Hutu militias. In some respects the committees followed the Rwandan model of ordered reconciliation (Clark 2014), but they underlined how different the context was from the Rwandan genocide and stressed the long history shared by Hutu and Tutsi in the region, which was confirmed by old men from both groups. The installation of the new and rebel-friendly mayor of Kiwanja, a small town north of Rutshuru, on the morning of 28 September 2012 was a typical example of the numerous public gatherings at which the M23 rebels failed to present themselves symbolically as an effective politicomilitary project supported by the government. Numerous high-ranking officers and politicians addressed an audience that consisted of a crowd of some 100 distracted children who had been forced to attend the event with their teachers as the rest of the population were seemingly not interested in the rebels' propaganda. One of the teachers summarized the attitude of the inhabitants thus: 'We don't like them and they know it. At the moment they are in power but they will disappear like they disappeared in 2009. We only hope that we do not have to suffer.'[8]

Institutional shape

This third aspect means institutionalized practices in administration, politics, economics, culture and education, which produce and perpetuate borders, symbols and their meaning.

The entire district administration was taken over and restructured by the rebels. In Rutshuru in September 2012, the M23's head of administration, Benjamin Mbonimpa, former deputy foreign policy spokesman of the CNDP, was supervising work being carried out along the main road. New signs were being erected with the slogan 'M23 – We are against corruption'. Mbonimpa told me: 'My dear Mr Martin, here we will show how a Congolese state could be. How things will function when all these incapable and corrupt liars have been replaced by competent and hard-working patriots.'[9]

Since the CNDP territory was mostly rural, the reactivation of agricultural advisory services was an important factor in creating territoriality. For example, the advisory services renegotiated the routes and times for moving cattle to new pastures, and fixed rates of compensation for fields destroyed by cattle. Legal disputes were dealt with by a local judiciary with new powers, similar to the traditional courts that used to exist in the domain of the Mwami, the 'king' of Bwisha, which roughly covered the area divided into two new administrative units by the political leaders of the CNDP. Police stations occupied by forces loyal to the CNDP or the M23 were instructed to strictly prosecute and prevent crime in order to deliver security:

I was in Rutshuru when the CNDP called me. And I came. I did not even hesitate. For the CNDP needs policemen who know their work... We had training and we were told that we had to be different from Kabila's police. Correct, not corrupt, straightforward and tough. Very tough. With everyone, do you understand?[10]

Prisons in Rutshuru and Bunagana filled with alleged thieves and bandits, as well as special military prisons for CNDP or M23 soldiers accused of attacking civilians, demonstrated the efforts of the rebels to take drastic measures against the frequently criticized lawlessness and impunity in eastern DRC. Although they appeared to be more disciplined than the FARDC, war crimes and breaches of human rights by rebel soldiers were one of the biggest credibility problems both for the CNDP and even more for the M23.

Highly symbolic was the takeover of the administration of Virunga National Park by the M23, including guided tours to the mountain gorillas for tourists. This was not only lucrative but was meant to show that the rebels were able to protect the gorillas, which attracted international attention. As a last example, the school system can be mentioned. Especially during the CNDP rebellion, schools stayed open, in contrast with many other schools in the rest of the province of North Kivu. On the one hand, this was due to the strike prohibition, which the rebels had imposed on the teachers. On the other hand, many young Congolese Tutsi from the Rwandan refugee camps volunteered as teachers. They got no salary but they were fed.

Effects and costs of the production of territoriality

What did the production of territoriality accomplish in the rebel areas and what price did the local people have to pay for it? During the CNDP rebellion, local peace, a strictly enforced monopoly on violence, the possibility of appealing to local lay courts, attention paid to local land-use conflicts and the work of a well-organized public service meant that the life of the rural population became more predictable. The CNDP, with its permanent and openly communicated authoritarian politicomilitary project, was a calculable risk, unlike the incalculable risks of a badly controlled government army with its polycephalous command structures. However, the end of unpunished violent attacks on the rural population by bands of unpaid government soldiers was purchased by means of new levies (money and/or foodstuffs) introduced by the CNDP, with which they financed the rebel army and the production of territoriality as described above. And although this calculability was recognized by some of the inhabitants of their territory,[11] at no time did the rebels have the approval or the active support of the majority of the people. Since the peasants had had to come to terms with many different violent actors over the previous 15 years, they developed a strategic relationship with the CNDP, which was often regarded as a mere puppet of the geopolitical interests of Rwanda. Forced recruitment, the effective prohibition of all political opposition and civil society protests, and not least financial losses suffered by border traders as a result of the war, were other serious consequences for the civilian population.

Apart from a few symbolic actions such as the anti-corruption campaign, very little of the M23 good-governance rhetoric materialized on the local level compared with the achievements of the CNDP. During the occupation of Goma in November 2012, the M23 rebels seized things that were useful for building 'their' state: computers and office equipment, road-construction equipment and lorries, money and, not least, great quantities of weapons and ammunition. Compared with the time of the CNDP, their means were fewer and they lacked the same political zeal. Their internal power struggles and their military opponents, who were well organized, unlike in the time of the CNDP, tied up their resources. The area controlled by the M23 thus did not become a territory in the sense described above but was more like an occupied zone whose constitution made life even more unpredictable than in the government-controlled areas. Due to the fighting between rival factions, the temporary abandonment of the district capital, Rutshuru, and violent attacks against civilians, there was no support worth mentioning among the local people. The lack of a charismatic leader with an authentic political vision, in the style of Laurent Nkunda, meant that the M23 was also uninteresting for the political opposition in the rest of the country.

Conclusions

This chapter has shown that in a situation of protracted civil war, new and fluid territorialities are constantly produced by rebel movements to confirm and legitimate their claims to power. The empirical vignette in the introduction, presenting the apparent paradox of an enclave of local peace, relative security and predictability, away from the media presence of the front and the refugee camps, has been explained through an analysis of the ephemeral politicoterritorial phenomena and the spatiotemporal structural dynamics of the war (Korf and Raeymaekers 2012).

More generally, the analysis can be read as a counternarrative that may help in questioning dominant assumptions about war in eastern DRC instead of intentionally or unintentionally justifying or legitimizing them. The rebels criticized the dysfunctionality of the Congolese state and derived from this the right to use violence. This was not an external challenge to state sovereignty based on the ascription of a 'weak' or 'failed' state along a Western ideal of the functionality of states (Hagmann and Péclard 2010). Rather, it was an endogenous diagnosis, based on years of experience of everyday violence, displacement and lawlessness. With the production of territoriality, the rebels worked on a territorially constituted countermodel, in the framework of which they practised their ideas of statehood.

That rebels produce security and fight for legitimate political causes even if they are supported by external powers is one of the many uncomfortable facts that are incompatible with the blueprints of, for example, the global state-centred regime of peacebuilding. This case study thus exposes a contradictoriness in concepts of war and peace that can also be found elsewhere and with which both scholars and practitioners have to deal. However, those actors engaged in conflict transformation are apparently obliged to actively ignore such complexity because it prevents them from applying the globally travelling scientific models dominant in this field. In contrast with Wiredu (2000), we might say that intended non-knowledge is necessary for action. In this regard I agree with Ravetz (1993), a philosopher of science who called the ignorance of ignorance a 'sin of science'.

After the disappearance of the CNDP, the local enclave of predictability turned out to be merely an intermezzo, and the defeat of the M23 showed how weak its political project was. However, this case study has revealed both successful and unsuccessful modes of the production of territoriality. It has shown that many different actors participate in the production of territoriality, such as soldiers, politicians, police officers, administrative officers and teachers. It is clear that the M23 wanted to continue the tradition of the CNDP, but that it lacked the necessary logistical, military and financial resources and political visions. A rebellion by Congolese Tutsi has seldom had so little support in its own local community and among the political opposition in the country. Consequently, the factor that made the CNDP strong – a territorially constituted rebellion with a standing army – led to the defeat of the M23. The group's legitimacy problem was aggravated by its failure to produce territoriality; its military defeat at the hands of MONUSCO and its Force Intervention Brigade was at least partly due to the fact that the rebels were simply not capable of controlling the vertical dimension and the volume of the territory (Elden 2013). In the end they had nothing with which to counter the attacks of a single new Rooivalk combat helicopter from South Africa.

Power and violence are not only exercised by the state and not only territorial. Rather, there is a contingent relationship between state, power, violence and territory. Neither state power nor non-state power must be constituted territorially, nor must they be exercised territorially.

But the well-known 'territorial trap' (Agnew 1994) should not mean ignoring processes of territorial production of violent orders and their reconfigurations (Newman 2010). This applies especially to political phenomena such as the territoriality of the rebellion that is described here. Embedded in parallel processes of de- and reterritorialization, in forms of political power in defined areas and cross-border flows and networks, as well as in interactions between abstract and material borders, between fluid and fixed territories, 'territory' is certainly not devoid of meaning both as an empirical object and as a theoretical construct. If it is not conceived as a fixed, rigid, immobile and unchangeable unit, but as a result of social practices such as those analysed in this case study; if it is empirically approached as an effect (Painter 2010), then territory is clearly of great analytical relevance to understanding the political structure of space and the spatial dimension of war and peace.

Notes

1. Interview, Bunagana, October 2008. All translations from French by me.
2. Telephone interview Gisenyi, Rwanda, March 2013.
3. Various informal talks, February and March 2014 and March 2015.
4. Interview in Jomba, October 2008.
5. Functionary of the CNDP, who worked for four years as a border official in Goma. Interview, Bunagana, October 2008.
6. Major Mbera Castro, in 2008 serving under Sultani Makenga, the CNDP commander of the so-called 'Bunagana Sector'.
7. Interview, October 2008.
8. Informal conversation, September 2012.
9. Interview, September 2012.
10. James, CNDP chief of police in Bunagana, interview, October 2008.
11. Various interviews, Bunagana, Jomba and Chengerero, October and November 2008.

References

Agnew J (1994) The Territorial Trap: The Geographical Assumptions of International Relations Theory. *Review of International Political Economy* 1(1): 53–80.

Agnew J (2005) Sovereignty Regimes: Territoriality and State Authority in Contemporary World Politics. *Annals of the Association of American Geographers* 95(2): 437–461.

Autesserre S (2010) *The Trouble with the Congo: Local Violence and the Failure of International Peacebuilding.* Cambridge: Cambridge University Press.

Autesserre S (2014) *Peaceland: Conflict Resolution and the Everyday Politics of International Intervention.* New York: Cambridge University Press.

Clark P (2014) Negotiating Reconciliation in Rwanda: Popular Challenges to the Official Discourse of Post-Genocide National Unity. *Journal of Intervention and Statebuilding* 8(4): 303–320.

Doevenspeck M (2011) Constructing the Border from Below: Narratives from the Congolese-Rwandan State Boundary. *Political Geography* 30(3): 129–142.

Elden S (2009) *Terror and Territory. The Spatial Extent of Sovereignty.* Minneapolis: University of Minnesota Press.

Elden S (2010) Land, Terrain, Territory. *Progress in Human Geography* 34(6): 799–817.

Elden S (2013) Secure the Volume: Vertical Geopolitics and the Depth of Power. *Political Geography* 34(2): 35–51.

Gross M (2010) *Ignorance and Surprise: Science, Society and Ecological Design.* Cambridge: MIT Press.

Hagmann T and Péclard D (2010) Negotiating Statehood: Dynamics of Power and Domination in Africa. *Development and Change* 41(4): 539–562.

High C, Kelly AH and Mair J (eds) (2012) *The Anthropology of Ignorance: An Ethnographic Approach.* New York: Palgrave Macmillan.

Knight DB (1982) Identity and Territory: Geographical Perspectives on Nationalism and Regionalism. *Annals of the Association of American Geographers* 72(4): 514–531.

Korf B, Engeler M and Hagmann T (2010) The Geography of Warscape. *Third World Quarterly* 31(3): 385–399.

Korf B and Raeymaekers T (2012) Geographie der Gewalt. *Geographische Rundschau* 64(2): 4–11.

McGoey L (2012) Strategic Unknowns: Towards a Sociology of Ignorance. *Economy and Society* 41(1): 1–16.

Newman D (2010) Territory, Compartments and Borders: Avoiding the Trap of the Territorial Trap. *Geopolitics* 15(4): 773–778.

Paasi A (1995) *Territories, Boundaries and Consciousness: The Changing Geographies of the Finnish-Russian Border.* Chichester: Wiley.

Paasi A (2003) Territory. In: Agnew J, Mitchell K and Toal G (eds) *A Companion to Political Geography.* Malden: Blackwell, pp. 109–122.

Painter J (2010) Rethinking Territory. *Antipode* 42(5): 1090–1118.

Prunier G (2009) *Africa's World War: Congo, the Rwandan Genocide, and the Making of a Continental Catastrophe.* Oxford: Oxford University Press.

Radil SM and Flint C (2013) Exiles and Arms: The Territorial Practices of State Making and War Diffusion in Post–Cold War Africa. *Territory, Politics, Governance* 1(2): 183–202.

Ravetz JR (1993) The Sin of Science: Ignorance of Ignorance. *Science Communication* 15(2): 157–165.

Sack RD (1986) *Human Territoriality: Its Theory and History.* Cambridge: Cambridge University Press.

Schlichte K (2009) *In the Shadow of Violence. The Politics of Armed Groups.* Frankfurt/Main: Campus.

Scott SA (2008) *Laurent Nkunda et la rébellion du Kivu. Au cœur de la guerre congolaise.* Paris: Karthala.

Sidaway J (2008) The Geography of Political Geography. In: Cox KR, Low M and Robinson J (eds) *The Sage Handbook of Political Geography.* London: Sage, pp. 41–55.

Stearns JK (2008) Laurent Nkunda and the National Congress for the Defence of the People (CNDP). In: Marysse S, Reynjens F and Vandeginste S (eds) *L'Afrique des Grands Lacs. Annuaire 2007–2008.* Paris: L'Harmattan, pp. 245–267.

Vlassenroot K and Raeymaekers T (2009) Briefing: Kivu's Intractable Security Conundrum. *African Affairs* 108: 475–484.

Wiredu K (2000) Our Problem of Knowledge: Brief Reflections on Knowledge and Development in Africa. In: Karp I and Masolo DA (eds) *African Philosophy as Cultural Inquiry.* Bloomington: Indiana University Press for the International African Institute, pp. 181–186.

3
Armed Conflict and Space: Exploring Urban–Rural Patterns of Violence

Kristine Höglund, Erik Melander, Margareta Sollenberg and Ralph Sundberg

Introduction

Where does large-scale violence take place? Large-scale collective violence is a multifaceted phenomenon and includes armed conflicts between government and organized opposition groups, violence between communal groups, and one-sided violence against civilians perpetrated by agents of the state or other armed actors. These forms of violence are usually studied as separate occurrences. In this chapter we are interested in the spatial dimension of these conflicts, exploring whether different forms of violence are predominantly rural or urban phenomena.

The motivation for the focus is two-fold. First, research concerned specifically with urban dimensions of violence has made important strides in recent years. For instance, it has sought to understand issues concerning urban warfare, and urban social unrest, where cities and urban centres are the central unit of analysis (see e.g. Buhaug and Urdal 2013; Graham 2010; Urdal and Hoelscher 2012; World Bank 2011). Yet we know that many conflicts are rooted in a rural context and that underdevelopment, marginalization and centre–periphery dynamics are core explanations for such violence (Bates 2008; Gurr 1970; Herbst 2000; Holsti 1996; Kalyvas 2006; Migdal 1988; Östby 2008; Scott 1977). These theories contain implicit or explicit ideas about the locus of violence. By treating rural and urban violence as distinct categories, but studying them jointly, it is possible to analyse the patterns of violence across the urban–rural divide; to explore to what extent these phenomena are the same or have different origins and dynamics; and to study the interlinkages between violence taking place in rural and urban areas.

Second, from a policy perspective it is important to retain a focus on both rural and urban patterns and sources of conflict. Urban populations are on the rise globally, and more than half of the world's population currently lives in urban areas (UN 2014). In many countries, the urban population makes up a disproportionately large share of the poor. A large part of this population ends up in slum settlements, where issues of governance are complicated by the informal nature of those settlements. These conditions provide fertile ground for intergroup tensions and social unrest. Yet a significant number of violent conflicts, not least in Africa, are primarily located outside the capital or urban areas. Eastern DRC, south Sudan and Darfur, northern Uganda and northern Nigeria provide only a few important examples. A full picture of large-scale collective violence therefore requires consideration of both rural and urban dimensions of conflict and unrest.

The chapter makes use of new data from the UCDP, hosted by the Department of Peace and Conflict Research at Uppsala University. We explore the spatial dimensions of conflict by making use of disaggregated and geocoded data covering all countries in Sub-Saharan Africa from 1989 to 2010. By using this spatially disaggregated information, we are able to move beyond the country level as the unit of analysis and study the exact location of violent events along the urban–rural dimension. Thus our study contributes to a recent wave of quantitative peace and conflict studies, which through the use of geographically disaggregated data has been able to explore the local and subnational patterns of violence (cf. Buhaug and Rød 2006; Fjelde and Hultman 2014; Raleigh and Hegre 2009; Wood 2014).

The chapter builds on, but extends, an initial analysis by Sundberg and Melander (2013) on the location of violent conflict. Their study outlines some of the patterns related to urban–rural dimensions of conflict. We take the analysis one step further by providing additional empirical analysis and more in-depth discussion of the implications for our understanding of rural versus urban conflict.

This chapter proceeds in the second section by briefly introducing previous research in relation to urban and rural dimensions of conflict. The third section presents the data used to determine the location and frequency of violence. In the fourth section, the main patterns of the location of violence are explored along different dimensions. We investigate if the absence or presence of violent localities varies between rural and urban areas, and if patterns concerning casualties are different in the rural and urban context. We also analyse whether trends of violence are similar or different across three categories of violence: state-based armed conflict, non-state armed conflict and one-sided violence. The patterns are also discussed in relation to the main theoretical arguments made in the literature on war, conflict and political violence. We conclude by proposing some implications for future research on the spatial dimensions of armed conflict.

The urban versus rural dimension of conflict

Spatial dimensions of conflict have been an important focus in peace and conflict research and have been studied both directly and indirectly. We know, for instance, which regions of the world have experienced the most armed conflicts in modern times. Conflict trends show that Asia and Africa are the regions with the overwhelming number of armed conflicts since 1945 (UCDP). Spatial dimensions in conflicts have also been examined through the contrasting focus lenses of urban versus rural. However, there is no comprehensive approach to studying the location of conflict from this perspective – both theoretical and empirical studies inform questions related to urban and rural violent conflict.

First, there is a strand of research which specifically focuses on violence and militarization in urban centres and cities. Given the global growth of urban populations – especially in the developing world – there has been an emergent interest in the city as a unit of analysis. To explore patterns and causes of urban social disorder, data on violence in cities have been compiled and analysed to study the influence of issues such as population growth and dispersion, economic development and growth, and globalization (Buhaug and Urdal 2013; Urdal and Hoelscher 2012).[1] In security studies, the urban context has been approached through issues of militarization and securitization of cities and urban populations. For instance, Graham (2010) explores the new military urbanism and the militarization of urban space. He points to how urban infrastructure becomes part of the security practices of the military, which in turn is a response to an urbanized perception of the enemy.

Second, several forms of violence contain explicit or implicit assumptions about the location of violence. For instance, terrorism – as one form of organized political violence – is by its very nature expected to occur more often in urban areas. To pursue a specific political goal, organized groups may use terrorist tactics aimed at putting pressure on governments. Terrorist tactics are 'designed to have far-reaching psychological repercussions beyond the immediate victim or target' (Hoffman 1998: 43). For this reason, the targeting of civilians in urban areas is a common tactic (Savitch 2001; Schmid 2011: 86–87). Given the high population densities in urban centres, urban terrorism is expected to maximize the desired effects. Similarly, coups and coup attempts, which can yield a substantial number of casualties, are initiated from within the state and directly target central power, which means that they almost always take place in the capital of a country where power is concentrated (see e.g. Luttwak 1968; Powell and Thyne 2011).

Third, research on political violence has also explored issues related to population density, which speaks to rural–urban dynamics, since urban areas tend to see a greater concentration of populations. A main argument in this literature is that capitals are likely to serve as battlegrounds and sites of contention because it is in capitals where stakes are concentrated. Areas become populated because they are resource rich or strategically located. These resources are important to control for governments and rebels alike. With control over a large part of the population as well as key resources, including means of communication and concentration of wealth, it also becomes easier to control the rest of the territory (Herbst 2000: 152–154). In addition, capitals and cities hold political and symbolic value, and therefore become important for rebels to control. Another argument proposes that population density may cause conflict because of the stress that large population concentration may inflict on the environment. In fact, problems of resource scarcity and land degradation may become more severe in areas with a high population density, such as urban centres (Raleigh and Urdal 2007). An alternative view suggests that marginalization and underdevelopment in areas far from the centre determine where the origin of rebellion is located. Ethnic and political grievances are often linked to geographical distance from the capital, making it possible for insurgents to mobilize the population and recruit members to their organization in such areas (Gurr 1970). Several propositions related to population density have been tested empirically by, for instance, Collier and Hoeffler (2004) and Raleigh and Hegre (2009). These studies find empirical support for more populated areas being more violence prone. Moreover, Raleigh and Hegre allude to urban–rural dynamics by showing that there is no clear-cut relationship between violent events and distance to capital. Their findings suggest that rural areas are important sites of conflict, in locations where there is also a large concentration of people (Raleigh and Hegre 2009: 237).

These different strands of research bring both complementary and conflicting findings which we will build on to provide a more nuanced understanding of different forms of political violence and its relation to urban–rural dynamics.

Exploring spatial patterns of conflict

In this chapter we explore urban–rural dimensions of conflict by studying the location of conflict events. Urban settings are understood as geographical areas where a large number of people cluster. The UN's *State of the World's Cities 2006/7* defines an urban area as the 'built-up or densely populated area containing the city proper, suburbs and continuously settled commuter areas' (UN 2006: 7).

To investigate where conflict takes place, we first establish a relevant unit of observation that can be assigned as rural or urban. We do this by constructing geographical units in Sub-Saharan Africa in the form of grid squares of 1 × 1 degree (c. 110 km × 110 km). The advantage of using this type of unit is that it is constant over time and space, and it is fully exogenous to factors associated with the onset of violence (cf. Tollefsen et al. 2012). An alternative would be to focus on some level of administrative unit. However, the relevant administrative unit varies greatly between different countries depending on their political system (e.g. centralized state versus federal state), which causes compatibility problems for cross-national comparison. Administrative units also change constantly, making it difficult to keep track of their exact borders at any given point in time. Moreover, the borders and makeup of administrative units change as a result of precisely those factors we believe may cause violence in the first place and would therefore be unsuitable for analysis since they are endogenous to conflict. Since this is the case, the geographical grid approach is judged to be more suitable for our purposes.

The squares employed in this study are relatively large compared with those used by other peace and conflict scholars. For example, Raleigh and Hegre (2009) employ 8.6 km × 8.6 km grid squares. We believe the larger units of observation are warranted for two primary reasons. First, if we are to capture the full dynamics of violence relating to urban centres as opposed to rural areas, we need squares that allow for the inclusion of whole cities. Second, recent analysis of disaggregated conflict data has shown that the reliability of reporting on violence drops dramatically with a resolution more fine-grained than 50 km × 50 km (Weidmann 2014). Thus larger units of observation are important in order to maintain an acceptable level of reliability of the information about violence.

After the grid squares were established, each was coded as urban or rural depending on whether it contains at least one urban location with a minimum of 100,000 inhabitants.[2] A grid square with at least one such urban centre is considered urban, and rural if there is no such urban location in the square (Sundberg and Melander 2013). Data for urban locations are taken from Nordpil and UN Population Division (2010) and from the Food and Agricultural Organization (FAO 2009).[3]

We explore the spatial dimensions of conflict by making use of disaggregated and geocoded conflict data covering all countries in Sub-Saharan Africa for 1989–2010. The data are taken from the UCDP Georeferenced Events Dataset. The dataset contains events within three categories of violence: state-based armed conflict, non-state armed conflict and one-sided violence. State-based conflicts are those that involve a state against either another state or a non-state actor in the form of an opposition organization. Non-state armed conflicts involve only non-state actors, which may be formally organized (typically pitting rebel organizations against each other) or informally organized (e.g. identity groups or supporters of political parties). One-sided violence involves instances where states or organizations target civilians. The information relating to each event includes 'date of the event, place of the event (with coordinates), actors participating in the event, estimates of fatalities, as well as variables that denote the certainty with which these data are known' (Sundberg et al. 2010).

To investigate the risk of violence in urban versus rural areas in Africa, we first establish whether there is any incidence of violence in the geographical units and whether that unit is urban or rural. We also establish the fatality levels in the geographical units. The units are observed for each year in the period 1989–2010, resulting in a total of 52,118 grid-year observations.

Patterns of violence versus no violence

A first reflection relates to the overall patterns of rural versus urban areas in Sub-Saharan Africa. As many as 88 per cent of all observations (grids) are considered to be rural. Although Sub-Saharan Africa is currently quickly urbanizing (UN 2014), this nevertheless corroborates the image of this region as being predominantly rural in the period covered by this study.

We also note that in only about 5 per cent of all observations, rural or urban, was an incidence of violence recorded. This implies that, on average, organized violence is a rare phenomenon where the majority of areas in Africa were spared such violence in the period 1989–2010. However, some countries, such as the Sudan, the DRC, Angola, Ethiopia and Somalia, with long-running, spatially dispersed and highly devastating civil wars, are clearly overrepresented in the data. For example, 16 per cent of all observations with violence were found in the Sudan, whereas 12 per cent were located in the DRC. In such war-afflicted states (e.g. Sudan, the DRC, Somalia and Liberia), all three types of violence are generally recorded, implying that different types of violence are linked and may give rise to, or feed on, each other. In a handful of countries – Benin, Gabon, Equatorial Guinea, Burkina Faso, Botswana and Cape Verde – no violence of the types covered by this study was recorded.

So how is violence dispersed over rural and urban locations? Table 3.1 displays the patterns across rural and urban geographical units.

A clear pattern emerges when looking at the occurrence of violence (any of the three types) in urban versus rural areas – the occurrence of violence in urban areas is almost triple that of violence in a rural area. This is consistent with population concentration arguments which hold that cities – and particularly the capital – are the main sites of contention because this is where key resources (e.g. material, political, strategic and symbolic) are concentrated, thus serving as strategic targets (cf. Herbst 2000; Lichbach 1995). This pattern is also consistent with previous empirical studies (e.g. Raleigh and Hegre 2009). There are numerous examples of the targeting of capital cities across Africa, including Monrovia (Liberia), Freetown (Sierra Leone), Bissau (Guinea-Bissau) and Mogadishu (Somalia), but also other strategic cities such as various provincial capitals in Angola targeted by UNITA (União Nacional para a Independência Total de Angola/the National Union for the Total Independence of Angola) and cities in eastern DRC.

Table 3.1 Violence versus no violence, 1989–2010

	Rural	**Urban**	**Total**
No violence	43,766	5,482	49,248
	95.50%	*87.13%*	*94.49%*
Violence	2,060	810	2,870
	4.50%	*12.87%*	*5.51%*
Total	45,826	6,292	52,118
	100%	*100%*	*100%*

Source: Sundberg and Melander (2013).

Patterns of different forms of violence

Previous research suggests that different forms of violence may be more or less connected to urban or rural dynamics. However, when analysing three different categories of violence separately, the same pattern in urban areas appears. Table 3.2 shows the patterns of violence in state-based conflicts.

When considering only state-based armed conflicts – that is, those that involve a state against another state or a non-state opposition organization – the risk of such conflict in an urban area is 2.5 times as high as in a rural area. The issue at stake in such conflicts is control over central government – or control over part of the state, if there are territorial claims. As a consequence, fighting often focuses on strategic urban targets at the centre or in the provinces. The localization of conflicts such as in Somalia, which has seen much of the fighting centred in and around Mogadishu and Kismayo, illustrates this point. Coups and attempted coups, a particular type of centralized state-based armed conflict, although relatively few for the period of study, are almost exclusively located in the capital. An example is the 1998 coup in Guinea-Bissau, which ultimately developed into civil war.

Are the patterns the same in non-state armed conflicts? Table 3.3 shows the patterns of violence in these conflicts.

Table 3.2 State-based armed conflict, 1989–2010

	Rural	**Urban**	**Total**
No violence	44,689	5,892	50,581
	97,52%	*93,64%*	*97,05%*
Violence	1,137	400	1,537
	2,48%	*6,36%*	*2,95%*
Total	45,826	6,292	52,118
	100%	*100%*	*100%*

Table 3.3 Non-state armed conflict, 1989–2010

	Rural	**Urban**	**Total**
No violence	45,316	6,039	51,355
	98.89%	*95.98%*	*98.54%*
Violence	510	253	763
	1.11%	*4.02%*	*1.46%*
Total	45,826	6,292	52,118
	100%	*100%*	*100%*

The category of non-state armed conflicts contains communal conflict, electoral violence between party supporters and conflict between competing rebel organizations. The latter often takes place in a larger context of civil war. The risk of non-state armed conflict is almost four times as high in urban areas as in rural areas. Communal conflict is sometimes assumed to be a predominantly rural phenomenon because it is often related to livelihood and land conflict in rural settings, such as herder–farmer conflict. However, communal conflicts are also common in urban areas, as witnessed in the long-running Hema–Lendu conflict in and around the city of Bunia, Ituri Province, in the DRC. Moreover, rebel organizations competing for power do so in both rural and urban areas. Again, Somalia and the DRC serve as examples of this.

The final category we explore is one-sided violence. Table 3.4 shows the patterns of violence across rural and urban locations in this type of violence.

The separate results for one-sided violence – that is, violence against civilians by states or non-state organizations – mirror those of the other types of violence. The risk of such violence in a given year and area is almost four times as high in urban areas compared with rural areas. In many states, violence against civilians occurred in connection to ongoing state-based armed conflict where the targeting of civilians can be conceived of as a strategy in civil war. The overall strategic aim of rebels targeting civilians would be to turn the population against the government, by showing that the government cannot protect them. For governments targeting civilians, the main aim would be to weaken the support base of the rebels and thus, ultimately, weaken the rebels (cf. Downes 2008; Hultman 2007, 2012; Kalyvas 2006). If violence against civilians is used as a strategy in war, it makes sense to see such instances in relation to urban areas – although not exclusively – where population and resources are concentrated and where violence can be expected

Table 3.4   One-sided violence, 1989–2010

	Rural	**Urban**	**Total**
No violence	44,935	5,824	50,759
	98.06%	*92.56%*	*97.39%*
Violence	891	468	1,359
	1.94%	*7.44%*	*2.61%*
Total	45,826	6,292	52,118
	100%	*100%*	*100%*

to have the maximum impact. Examples of this include the deliberate targeting of civilians in urban areas in Liberia, Sierra Leone, the DRC, Rwanda and Burundi.

Patterns of intensity of conflict and location

To obtain a more nuanced picture of violence in urban and rural areas, we explore if there are any differences in the intensity of rural or urban violence. Specifically, we look at the number of casualties in rural and urban conflict locations. The number of casualties across these areas is shown in Table 3.5.

Whereas violence is more likely in urban areas, the available data show that rural conflict is in fact more violent. Almost half of all coded deaths occurred in rural areas. Separating into the different categories of violence, the same pattern is visible for state-based armed conflict as well as non-state armed conflict. Roughly two-thirds of all deaths in state-based armed conflict occurred in rural areas, and the same overall pattern applies to non-state armed conflict where about 60 per cent of the deaths took place in rural areas. However, the proportions are reversed in the case of one-sided violence, where two-thirds of the deaths were incurred in urban areas.

Looking closer at what these figures contain, it is clear that many of the deaths in state-based armed conflicts were associated with particular conflicts that may have predominantly rural or urban characteristics. The interstate armed conflict between Ethiopia and Eritrea, which was fought in a largely unpopulated area and where fighting resulted in almost 100,000 deaths, accounts for part of the association between state-based armed conflict and deaths in rural areas. However, in general, the large number of deaths in state-based armed conflicts associated

Table 3.5 Casualties in urban versus rural areas, all violence and per category, 1989–2010

	Rural	Urban	Total
All violence	362,975	366,515	729,490
	49.76%	*50.24%*	*100%*
State-based conflict	227,448	137,030	364,478
	62.40%	*37.60%*	*100%*
Non-state conflict	40,868	27,412	68,280
	59.85%	*40.15%*	*100%*
One-sided violence	94,659	202,073	296,732
	31.90%	*68.10%*	*100%*

with rural areas compared with urban areas may be a function of the nature of warfare during this period. Rebels have had relative advantages *vis-à-vis* the state – which in several weak African states has mainly maintained a strong presence in the capital or major cities – in less populated non-urban areas and thus have inflicted relatively higher death tolls on each other. This pattern is still compatible with a higher risk of violence in urban areas in a given year because urban areas constitute important targets for rebels aiming to take government control and are thus constantly targeted.

For non-state armed conflicts, large numbers of deaths in non-state armed conflict were recorded in countries such as the DRC, where rebel groups and militias have been active in non-urban areas in which the government has had little control or even presence. The same pattern can be seen in the fighting between factions of the Sudan People's Liberation Army in Sudan in the early 1990s, as well as in Somalia after the fall of Siad Barre in 1991. In this type of weak state, state presence is at best linked to major cities, and areas outside state control are predominantly rural. With no state presence in peripheral areas and little presence of other actors who could potentially mitigate violence, non-state groups are often violently vying for control (Fjelde and Nilsson 2012). In these settings, fighting can continue unchecked and thus result in large death tolls. Moreover, when the government is biased against one side, as has been the case in communal conflicts such as those in Darfur, violence also becomes more intense (Brosché 2014). In other words, the absence of the state as well as the involvement of a biased state in these predominantly rural areas can fuel conflict intensity and thus raise casualty levels in non-state conflicts.

Unlike state-based armed conflict and non-state armed conflict, the majority of deaths in one-sided violence were recorded in urban areas. This is to some extent explained by the genocide in Rwanda in 1994 where most deaths are recorded in grid squares coded as urban areas. Large numbers of deaths were inflicted in and around Kigali at the beginning of the genocide. During the years following the genocide, more large-scale massacres occurred in urban areas. There were also a number of large-scale killings of civilians in, for example, Monrovia, Liberia, in 1990 as the National Patriotic Front of Liberia first entered the capital targeting civilians associated with President Doe, as well as in mid-1996 in the final stages of the war. Other examples include several large-scale massacres of civilians in 1996 in eastern DRC, near Goma and Bukavu (North and South Kivu, respectively) at the very beginning of the war in the DRC. In all of these examples, violence against civilians occurred in connection with ongoing state-based armed conflict and with key developments in those wars (cf. Hultman 2007, 2012; Kalyvas 2006). It should also be noted that few events in this period can be classified as terrorism defined in the traditional sense.[4] However, the end of this time period witnessed the rise of groups such as al-Shabaab, active in Somalia and neighbouring countries, as well as al-Qaeda in the Islamic Maghreb, active in the Sahel region. Increasing numbers of violent terrorist acts have been staged by such terrorist groups in recent years, targeting urban as well as rural areas.

Discussion: Additional observations, confounders and systematic biases

This study has generated several important insights into the prevalence of violence in urban versus rural areas. However, the potential and the limitations of these findings warrant some further discussion.

First, our analysis of urban versus rural patterns of violence has not explored whether less populated rural areas display different features from rural areas with a high population density. A cursory analysis suggests that the impact of population density is different in these contexts (Sundberg and Melander 2013). More populated areas are found to be more violent. It can be argued that for rural areas it is only when such regions are sufficiently densely populated that violence will arise. People do the fighting and conduct the violence, and there must be a minimum level of population for people to join together to organize for violence (Lichbach 1995). Moreover, during – or in connection with – civil wars, we may expect fighting when there are resources to fight over, which often revolves around a population that serves as a resource for the party in terms of recruitment, supplies or intelligence, and which, conversely, may be associated with the other side along the same dimensions (Lichbach 1995; Kalyvas 2006). Prolonged fighting is rare in areas where there is a low, or zero, population. A notable exception is the interstate war between Ethiopia and Eritrea in 1998–2000. In that case, armies moved into the border area of contestation around Badme, an area which in itself had very few and only minor population centres. For urban areas, the prerequisites for organization for violence (i.e. large numbers of people and various types of resource) are already fulfilled, and thus no similar pattern between violence and population density can be discerned. Expressed differently, there are diminishing returns of population density in urban areas, but not in rural areas.

Second, in our analysis it is clear that some individual cases of armed conflict drive the results and are very important in an interpretation of the patterns which emerge in the African context. Deaths from the Rwandan genocide, which was to a large extent associated with urban areas, constitute a large portion of all deaths in one-sided violence. Also, the war between Ethiopia and Eritrea which was fought in a non-urban context is responsible for a large number of deaths in the category of state-based armed conflict. However, it is important to note that these cases do not seem to alter the main patterns but only to reinforce them. Even when these cases are excluded, casualty numbers still seem to be higher in rural areas for state-based and non-state armed conflicts, but not for one-sided violence where the largest numbers of deaths are associated with urban areas.

Third, an analysis of urban and rural violence needs to consider different dynamics in small countries versus large countries. In order to fully capture urban–rural dynamics in small countries such as Rwanda, Burundi, Liberia and Sierra Leone, it would be necessary to disaggregate the analysis even further by, for instance, using smaller units of observation. In order to take differences between countries into account – in terms of size or any other relevant aspect – an alternative would be to use more sensitized measures of what constitutes an urban or a rural area. However, such measures are currently not available for all of Sub-Saharan Africa and require the resource-intensive coding of urban and rural areas for each country.

Finally, an urban bias in reporting of violence creates obstacles to how far we can take the analysis (Galtung and Ruge 1965; Kalyvas 2006: 41; Weidmann 2014). There are several dimensions related to this issue, including the higher costs (and sometimes risks) involved in gathering data in rural areas. For the types of source this study relies on – that is, mainly news sources – we have good reason to assume that there will be an underreporting on events, and that this underreporting increases with remoteness/low population density because of the number of potential observers. Underreporting in rural areas may also be a result of urban news being more attractive for news media to report on, or a lack of capacity in media organizations to cover events in rural settings. However, the extent of the problem for our analysis is difficult to assess because it is impossible to verify which events were not reported. What we do know is that types of error in reporting are not uniform. For instance, Weidmann (2014) finds that news reporting on casualty rates is generally accurate even if these are incurred in remote areas, whereas the location is more often inaccurately reported. At this stage we cannot fully assess the extent to which our finding of urban areas being more prone to violence is partially a result of a systematic underreporting in rural areas. It can be argued that the likelihood of underreporting is greater for one-sided violence as well as for nonstate conflict than for state-based conflict, where both sides have more channels to publicize their actions. There is also great variation in the accuracy of reporting between countries depending on the level of government monopoly of information, the level of independent observer presence and the resources available to the actors (and victims) for transmitting information.

Concluding remarks

Armed conflict in Africa is multilayered, and its causes and effects require careful analysis. This chapter has explored where different forms of violence take place and if armed conflicts are predominately rural or urban. Our approach of employing disaggregated subnational-level data on violence, yet doing this for a large region (Sub-Saharan Africa) and for an extended time period (1989–2010), allows us to establish general patterns in regard to rural–urban aspects of violence. Aside from its descriptive merits, patterns derived from such a spatially disaggregated approach can be used for comparisons with implications for existing theories, as well as for refining and developing theory. Our analysis suggests three main conclusions based on a selection of distinctive patterns found in this study, and which future research concerned with space and conflict dynamics may investigate further.

First, large-scale collective violence is a rare phenomenon, and our disaggregated analysis shows that few countries and few areas account for a large share of violence. These countries include Sudan, the DRC, Somalia and Liberia, which also experienced state-based, non-state and one-sided violence. Our analysis thus supports the notion that different forms of conflict are interconnected. Of importance for this chapter is that these countries also experienced substantial rural as well as urban violence, indicating that violence in different types of areas may be interlinked. This implies an integrated approach where violence should not be studied from an exclusively rural or urban perspective.

Second, our analysis shows that urban locations experience armed conflict more often than rural locations. This is true regardless of the type of violence explored in this chapter. While part of this finding may be related to an urban bias in the reporting of violence, there are also good theoretical reasons for believing that urban areas are particularly vulnerable to violence, given their high population density and their economic and symbolic significance. This suggests a continued focus in peace and conflict research on the specific predicaments of social unrest and large-scale political violence in urban areas.

Third, while urban areas are more prone to experiencing violence, rural locations have a larger share of casualties. This trend is clearly visible for state-based and non-state conflict, including civil wars and communal conflict. This means that peace and conflict research needs to retain a focus on the rural dimensions of armed conflict, since violence in rural areas, despite occurring less frequently than in urban areas, has considerable consequences for the societies in which it takes place. Moreover, although violence often involves urban areas, the origins of conflict in terms of underdevelopment and marginalization may still be based in conditions in rural areas, which, in turn, risk being reinforced by the effects of violence in these vulnerable areas.

Notes

1. For instance, the dataset used by Urdal and Hoelscher covers 55 major cities in Asia and Sub-Saharan Africa for 1960–2009 and contains 'nonviolent actions such as demonstrations and strikes and violent political actions like riots, terrorism and armed conflict' (2012: 512).
2. Note here that refugee settlements are not included. In terms of population size, some refugee camps, notably those housing Somali refugees in Kenya (e.g. Hagadera, Dagahaley), fulfil the population criterion for cities used in this study. Although theoretical arguments focusing on how population density may lead to violence could potentially apply to refugee camps as well as to cities, we nevertheless consider them to be separate phenomena due to their highly particular characteristics. We therefore exclude refugee settlements from our study of the urban–rural dimensions of violence.
3. There are a number of alternative measures and data for the urban–rural distinction. Aside from the data used in this study, population data are provided by the Socioeconomic Data and Applications Center at the Center for International Earth Science Information Network, which holds a number of different datasets on both population and urban areas. Other examples include UN Habitat's (2008) Global Urban Observatory data. However, the latter have been found to be problematic due to missing data (Urdal and Hoelscher 2012). An alternative measure which has been used in recent years is data on night light emission. However, this speaks primarily to levels of wealth and poverty rather than to the urban–rural distinction *per se* (Chen and Nordhaus 2011; Shortland et al. 2013).
4. By 'terrorism' we mean violence by politically motivated non-state actors which targets civilians with the purpose of distilling fear in a larger audience than the immediate victims (cf. START 2014). Some authors argue that this also includes civilian targeting by groups such as UNITA (União Nacional para

a Independência Total de Angola/National Union for the Total Independence of Angola) and RENAMO (Resistência Nacional Moçambicana/Mozambican National Resistance), as well as the LRA (Lord's Resistance Army) in Uganda and LURD (Liberians United for Reconciliation and Democracy) (Cilliers 2003). This would mean a very different interpretation of the number of terrorist events in the period covered by this study.

References

Bates R (2008) *When Things Fell Apart*. Cambridge: Cambridge University Press.

Brosché, J (2014) *Masters of War: The Role of Elites in Sudan's Communal Conflicts*. Uppsala: Department of Peace and Conflict Research, Uppsala University.

Buhaug, H and Rød, JK (2006) Local Determinants of African Civil Wars, 1970–2001. *Political Geography* 25(3): 315–335.

Buhaug H and Urdal H (2013) An Urbanization Bomb? Population Growth and Social Disorder in Cities. *Global Environmental Change* 23(1): 1–10.

Chen X and Nordhaus WD (2011) Using Luminosity Data as a Proxy for Economic Statistics. *Proceedings of the National Academy of Sciences* 108(21): 8589–8594.

Cilliers J (2003) Terrorism and Africa. *African Security Review* 12(4): 91–103.

Collier P and Hoeffler A (2004) Greed and Grievance in Civil War. *Oxford Economic Papers* 56(4): 563–595.

Downes AB (2008) *Targeting Civilians in Civil War*. Ithaka: Cornell University Press.

FAO (2009) City Location and Population in Africa, 2009 version. Available at: http://www.fao.org/geonetwork/srv/en/main.home (accessed 12 April 2015).

Fjelde H and Hultman L (2014) Weakening the Enemy: A Disaggregated Study of Violence against Civilians in Africa. *Journal of Conflict Resolution* 58(7): 1230–1257.

Fjelde H and Nilsson D (2012) Rebels against Rebels: Explaining Violence between Rebel Groups. *Journal of Conflict Resolution* 56(4): 604–628.

Galtung J and Ruge MH (1965) The Structure of Foreign News: The Presentation of the Congo, Cuba and Cyprus Crises in Four Norwegian Newspapers. *Journal of Peace Research* 2(1): 64–90.

Graham S (2010) *Cities under Siege: The New Military Urbanism*. London: Verso.

Gurr TR (1970) *Why Men Rebel*. Princeton: Princeton University Press.

Herbst J (2000) *States and Power in Africa. Comparative Lessons in Authority and Control*. Princeton: Princeton University Press.

Hoffman B (1998) *Inside Terrorism*. London: Indigo.

Holsti K J (1996) *The State, War, and the State of War*. Cambridge: Cambridge University Press.

Hultman L (2007) Battle Losses and Rebel Violence: Raising the costs for fighting. *Terrorism and Political Violence* 19(2): 205–222.

Hultman L (2012) Attacks on Civilians in Civil War: Targeting the Achilles Heel of Democratic Governments. *International Interactions* 38(2): 164–181.

Kalyvas SN (2006) *The Logic of Violence in Civil War*. Cambridge: Cambridge University Press.

Lichbach M (1995) *The Rebel's Dilemma*. Ann Arbor: Michigan University Press.

Luttwak E (1968) *Coup d'État. A Practical Handbook*. London: Penguin.

Migdal JS (1988) *Strong States and Weak Societies. State-Society Relations and State Capabilities in the Third World*. Princeton: Princeton University Press.

Nordpil and UN Population Division (2010) World Database of Large Urban Areas, 1950–2050, version 1.1. Available at: http://nordpil.com/go/resources/world-database-of-large-cities/ (accessed 12 April 2015).

Östby G (2008) Polarization, Horizontal Inequalities and Violent Civil Conflict. *Journal of Peace Research* 45(2): 143–162.

Powell JM and Thyne CL (2011) Global Instances of Coups from 1950 to 2010: A New Dataset. *Journal of Peace Research* 48(2): 249–259.

Raleigh C and Urdal H (2007) Climate Change, Environmental Degradation and Armed Conflict. *Political Geography* 26(6): 674–694.

Raleigh C and Hegre H (2009) Population Size, Concentration, and Civil War. A Geographically Disaggregated Analysis. *Political Geography* 28(4): 224–238.

Savitch HV (2001) Does Terror Have an Urban Future? *Urban Studies* 38(13): 2515–2533.

Schmid AB Ed (2011) *Handbook of Terrorism Research*. London: Routledge.

Scott JC (1977) *The Moral Economy of the Peasant. Rebellion and Subsistence in Southeast Asia*. New Haven: Yale University Press.

Shortland A, Christopoulou K and Makatsoris C (2013) War and Famine, Peace and Light? The Economic Dynamics of Conflict in Somalia 1993–2009. *Journal of Peace Research* 50(5): 545–561.

START (2014) Global Terrorism Database Codebook: Inclusion Criteria and Variables. National Consortium for the Study of Terrorism and Responses to Terrorism (START), College Park, MD. Available at: http://www.start.umd.edu/gtd/downloads/Codebook.pdf (accessed 12 April 2015).

Sundberg R, Lindgren M and Padskocimaite A (2010) *UCDP GED Codebook Version 1.0-2011*. Department of Peace and Conflict Research: Uppsala University.

Sundberg R and Melander E (2013) Introducing the UCDP Georeferenced Event Dataset. *Journal of Peace Research* 50(4): 523–532.

Tollefsen AF, Strand H and Buhaug H (2012) PRIO-GRID. A Unified Spatial Data Structure. *Journal of Peace Research* 49(2): 363–374.

UN (2006) *State of the World's Cities 2006/7*. Nairobi: UN Human Settlements Programme (UN-HABITAT).

UN (2014) *World Urbanization Prospects: 2014 Revision, Highlights*. New York: Department of Economic and Social Affairs, UN.

Urdal H and Hoelscher K (2012) Explaining Urban Social Disorder and Violence: An Empirical Study of Event Data from Asian and Sub-Saharan African Cities. *International Interactions* 38(4): 512–528.

Weidmann N (2014) On the Accuracy of Media-based Conflict Event Data. *Journal of Conflict Resolution*, Epub ahead of print 29 April 2014, DOI: 10.1177/0022002714530431.

Wood RM (2014) From Loss to Looting? Battlefield Costs and Rebel Incentives for Violence. *International Organization* 68(4): 979–999.

World Bank (2011) *Violence in the City*. Washington, DC: World Bank.

Part II
Global and Local

4
Reading Urban Landscapes of War and Peace: The Case of Goma, DRC

Karen Büscher

Introduction

Over the past 20 years the city of Goma, the administrative capital of the Congolese North Kivu Province located at the Congo–Rwanda border, has become a regional urban symbol of violent conflict dynamics in eastern DRC, as well as of peacebuilding and post-conflict reconstruction. The city's position in the conflict-ridden Kivu, its economic significance in the regional political economy of war and its security position as a destination of internally displaced persons (IDPs) as well as hundreds of humanitarian, development aid and peacebuilding agencies all make Goma a point of 'central marginality'. Situated in the borderlands far from the national political centre at the violent margins of the state, Goma developed into a booming economic and humanitarian hub with a strong political and military position in the Great Lakes Region (Vlassenroot and Büscher 2013).

As a central location for both conflict and peace processes, Goma has offered a fascinating research site for political scientists, anthropologists, human geographers and conflict researchers to investigate the local impact of political dynamics related to the Kivu crisis, such as the militarization of governance (Büscher 2012; Mampilly 2011; Tull 2005; Verweijen 2013), the political economy of war (Jackson 2002, 2003; Lamarque 2014; Laudati 2013; Raeymaekers 2010; Titeca 2012), the violent mobilization of ethic identities (Boas and Dunn 2014; Dunn 2001; Mararo 2002; Vlassenroot and Büscher 2013) forced displacement economies (Rohwerder 2013), the humanitarian industry (Büscher and Vlassenroot 2010), the local politics of international peacebuilding efforts and security reforms (Autesserre 2010; Tull 2009) and the adaptation of urban livelihoods to a situation of violent uncertainty and risk (Büscher and Mathys 2013; Oldenburg 2010).

In the light of the so-called 'spatial turn' of these political and social sciences, some of these studies have explicitly taken the city as the particular spatial urban unit from which to explore regional sociopolitical dynamics related to the processes of war and peace (Büscher 2011; Doevenspeck 2013; Oldenburg 2010; Tull 2005, 2009). From a political geography approach, the dynamics of violent conflict in the Kivu region have been studied extensively in relation to spatial categories such as geographic location, distribution of resources and territorial control. Connecting the logics of conflict and violence in eastern DRC to struggles over land, to the complex political history of the Congo–Rwanda border, to the presence of conflict minerals, or to post-colonial struggles over territorial identity and 'autochthony' and so forth are all examples of a 'geographies of war' approach (Autesserre 2008; Doevenspeck 2011; Eichstaedt 2011; Jackson 2006a, 2006b; Le Billon 2001, 2013; Van Acker 2005; Vlassenroot and Büscher 2013).

Within these approaches, which represent a very valuable contribution to a better understanding of the recent political and socioeconomic conflict history of this region, it is in particular the borderland approach that fully demonstrates the reciprocity of the spatial dimension of violent conflict. Studies considering Goma as an urban borderland (a distinctive geographic, political and social space, producing particular urban livelihoods, modes of governance, cultures and identities) have clearly shown how on the one hand spatial constellations produce political and social action, and on the other hand how this action is producing and (re)structuring the urban space (Büscher and Mathys 2013; Doevenspeck 2011, 2013, Chapter 2 in this volume; Lamarque 2014).

This chapter aims to take the 'spatial reading' of the transformation of Goma into a booming regional political and economic centre, the headquarters of violent conflict as well as of peacebuilding, to another level. With particular attention being paid to the interaction between the production and reproduction of the physical urban landscape and dynamics or spatial 'practices' of conflict and peace, it contributes to the study of the spatial situatedness of Goma's contemporary identity as both a 'city of rebellion' and an 'urban peaceland'.

It is obvious that 20 years of violence, militarization, forced displacement and humanitarian crisis had a strong impact on Goma's cityscape. The informal and conflictual urbanization process of the city is the spatial outcome of its complex political trajectory; in the urban landscape one can read both the local effects of civil war and the localized dynamics of peacebuilding. The changing use and function of the urban space by its elites and 'ordinary' dwellers, the transformation of public space into military camps or humanitarian offices, the conversion of refugee camps into residential areas and the increased contestation between urban military elites over neighbourhoods, streets or roundabouts are all features of the dynamic relationship between 'conflict-and-peace' processes and processes of spatial reconfiguration.

Both peace and conflict take place in Goma, simultaneously. Goma is a clear illustration of a complex 'no-war-no-peace' reality (Dijkzeul 2008; Larmer et al. 2013; Swart 2011; Themnér 2011) where, despite the signing of yet another peace accord, societies continue to be militarized and persisting violence and insecurity continuous to produce precarious urban life conditions. In Goma this no-war-no-peace situation is palpable from the levels of urban governance to everyday livelihoods. Urban elites, connected to rebel leaders and local politicians, play an important role in urban governance. Violence in the rural hinterlands constantly pushes IDPs into the city. Urban university students are actively recruited to join rebel movements. The same students are applying for jobs in the international donor community, humanitarian sector or peacebuilding community that has become a major employer in town. Young entrepreneurs invest their money in conflict minerals and in expat real-estate projects at the same time.[1] This chapter will start from the translation of this complex no-war-no-peace reality and move to the use, transformation and (re)production of urban landscape.

The first section will investigate in further detail this dual urban identity. Both dominant characteristics of Goma as a 'city of rebellion' and an 'urban peaceland' will be historically contextualized. Further, on the basis of empirical evidence it will be demonstrated how these characteristics are translated not only into local socioeconomic and political constellations but more importantly for this chapter into material urban landscapes.

In the second section I move beyond this dual urban identity by focusing on active spatial engagement, practices and agency from within by different kinds of urban actors. After introducing the particular context of Goma's urbanization process in a situation of state failure and informality, I investigate, as a case study, the spatial strategies of public authority by local urban 'big man'. It will be argued that the eventual urban space that is being generated from within, instead of reproducing dynamics of conflict and peace, produce an alternative image of Goma, constructed around its identity as a city of leisure, tourism, prosperity and opportunity.

By emphasizing local and international actors' agency in practices of war and peace and beyond, and how these are being materialized through the use and transformation of the urban space, this chapter aims to achieve a better understanding of the dynamic production of space in an urban 'no-war-no-peace' zone. As such, the spatial approach to dynamics of conflict and peace used here starts from the academic perspective of political anthropology or political ethnography, taking local practices, strategies, identities and livelihoods as the starting-point of the analysis.

This chapter is built on original empirical ethnographic data collected by me during several research stays in Goma, North Kivu, in eastern DRC between 2008 and 2014. This doctoral and postdoctoral field research has been financed by Ghent University and the Flemish Research Foundation. Mostly ethnographic research methods were used, such as participant observation, semistructured interviews and focus-group discussions with different urban stakeholders.

Goma is a provincial capital city of an estimated 900,000 inhabitants with a cosmopolitan ethnic background. It experienced its most important expansion from the end of the 1990s onwards in the Second Congo War in the context of a natural resources-driven war economy. More recent waves of violence have pushed many IDPs into the city, leading to further, largely uncontrolled geographic and demographic expansion. Goma is located right at the border with Rwanda, on the shores of the Kivu Lake and on the foot of Nyiragongo Volcano.

'No-war-no-peace' societies as an analytical framework

In this chapter I approach this case study – the city of Goma – as an urban no-war-no-peace society to better conceptualize war and peace taking place at the same time and equally shaping and reshaping the city. Richards (2005) introduced the concept of 'no-war-no-peace' as a basis for a better analytical understanding of the complex violent conflicts emerging from the post-Cold War setting. Based on case studies such as Cambodia, Somalia, Bosnia and Uganda, he demonstrated that one has to go beyond the sharp categorical distinction between war and peace and think more in terms of a 'continuum' to grasp the complex effects of current conflict dynamics on local societies. A situation of a constant threat of re-emerging violence, states and security forces unable to secure their citizens, the high mobility of the population, the hybrid nature of actors involved in governance, the strong militarization of society and so on are some examples of characteristics of a no-war-no-peace society, which are perfectly applicable to the case of Goma.

A number of scholars followed Richards in attempts to further analytically and conceptually expand and valorize the no-war-no-peace concept (Dijkzeul 2008; Mac Ginty 2006; Yanacopulos and Hanlon 2006). A particularly valuable outcome of these studies is the understanding of situations of protracted violent conflict, despite their contextual specificities, as a distinct social reality, or social setting, generating distinct social characteristics. The 'urban' as a particular analytical starting-point is very interesting in this regard since it tends to 'magnify' the dynamic interactions between the processes of war and peace and their local effects (Beall et al. 2011; Büscher 2011). Cities function as concentrations of productions and practices of war and peace initiatives; the spatial, political and socioeconomic outcomes of the no-war-no-peace setting are particularly visible there. As cities are 'laboratories of change' (Robinson 2006), they form a perfect site from which to study the transformative effects of a protracted no-war-no-peace situation. For this chapter what I take as particularly useful from the no-war-no-peace framework is the emphasis on these social settings as productive places. Endorsed by the literature on the transformative power of violence and war (Cramer 2006; Duffield 2001; Keen 1998), I think it is important to look beyond the mere destructive effects of protracted conflict to also underline the productive outcomes (of war *and* peace, as they both equally shape Goma's current identity). No-war-no-peace societies form social spaces that produce and reproduce particular modes of governance, local economies, political and social orders, institutions, identities, spaces, cultures, lifestyles and so on (Lambach 2007; Richards 2005). It is precisely in that frame that we have to understand spatial practices and spatial agency in Goma. Building on the recommendation of Richards to study the practices of war and peace, this chapter will investigate the simultaneous spatial strategies, actions and productions of war and peace by different groups of urban actors.

Goma, the 'city of rebellion': A city where conflict takes place

Through recent history the city of Goma has increasingly been perceived from outside as a zone of war, a centre of insurgency, and a city of chaos, violence and insecurity. Situated at the foot of Nyiragongo Volcano (one of the most active volcanoes in Africa), Goma is also believed to be 'volcanic' in another sense (Mpisi 2008) – referring to the city as a 'powder keg', at the heart of what one calls 'the soft belly of Congo's security situation' (Pole Institute 2007). Its central position in the different wars that have been afflicting eastern DRC has made it the urban focal point of rebellion.

Its significance as the city of rebellion was given a literal interpretation when Goma became the headquarters of the RCD[2] rebels in 1998, which turned it into a highly militarized 'city-state'. However, after the peace agreements (2004), the reunification of the country and the democratic elections (2006), Goma did not get rid of its sinister reputation. Since 2007, during the rebellions of the CNDP and the M23, the city was targeted several times, illustrating its strategic and symbolic meaning within the regional armed conflict. For the rebel forces, taking over the city was an important political demonstration of their power and control (Büscher 2012; Oldenburg 2010; Berwouts 2012).

At present the picture of Goma as a rebel city has a layered significance. It points at the visible militarization of the city, the prominent presence and circulation of arms, the dominant behaviour of armed actors, the increasing urban violence and the 'insurgency' identity of the inhabitants (Büscher 2011). The urban 'rebel' character thus has to be interpreted as a practice, identity and a form of urbanity.

Goma is not a place where 'conflict takes place' in the sense that it has itself become an urban battlefield where different armed groups violently clash with each other as they did in Bangui, Monrovia or other African urban 'warscapes'. Illustrations of how violent conflict manifests itself are, for example, the way this urban space is governed, the particular characteristics of the main actors involved in urban governance, the function and use of violence in these actors' urban strategies of power, access, authority and control, the large presence of security forces and in the military connections of the local urban economy.

A situation of protracted violent conflict in a context of a weakened Congolese state created an urbanization process that was marked by a profound informalization (with regard to housing and other public services, employment and administration) and a strong (sometimes violent) fragmentation of the urban space. Protracted civil war has deepened existing fault lines at the urban level, reinforcing the fierce local contestations (over resources, land, economic opportunities and citizenship), often along ethnic lines (Büscher 2012).

These developments have resulted in a spectacular spatial urban transformation. The city is one of the fastest urbanizing centres in the country and the region (De Saint Moulin 2010). Its exceptional demographic and geographic expansion is directly related to the context of war and insecurity; a steady influx of IDPs into the peripheral parts of town has put the city under strong pressure (Büscher 2011; Verhoeve 2004). During the most recent period of intensified violence in November 2013, some 150,000 IDPs were pushed into the city (IRIN Africa 2012), resulting in the telling observation that 'Goma was running out of space' (IRIN Africa 2013). The IDP camps on the outskirts as informal, temporary urban neighbourhoods have become an important symbol of the urban conflict landscape of the town.

Apart from these IDP camps, other particular transformations of Goma's urban landscape are locally referred to as markers of the city's identity as a 'rebel city'. The increasing spatial occupation by armed 'security' actors in town is one example. The increased militarization of urban life has materialized in the form of an overwhelming presence of military staff, police forces, private security companies, an impressive UN peacekeeping force (MONUSCO) and informally operating urban self-defence units. Military barracks, UN compounds and circulating tanks with blue helmets all have become dominant features of the urban landscape. Military actors also mark private space in Goma; urban inhabitants secure themselves by hiring a soldier from the FARDC as a private guard, and influential (ex-)colonels and generals own large parts of real estate in the wealthy neighbourhoods of town (*Goméens* can easily identify which building belongs to which colonel from which armed group).

A recent allocation of one of the last remaining strips of open public space in the city centre (along the shores of Kivu Lake) to the colonels of the FARDC as a reward for their military efforts in defeating the M23 is another illustration of the military appropriation of public space and transformation of urban landscapes.

A last straightforward effect of the dynamics of protracted conflict on Goma's urban transformation is the development of the 'international humanitarian sector'. The number of international development and humanitarian agencies that came to settle in Goma has grown steadily since their first arrival in response to the Congo–Rwanda refugee crisis in 1994. While most of their programmes and actions focus on the rural hinterlands of the city, they all have their main office in town, where their international staff reside. The presence of this 'humanitarian sector' as an effect of ongoing insecurity and crisis has become an integral part of the city's current identity and landscape, and it is extremely visible when circulating in town: the international and local urban NGO landscape marked by the many white sport utility vehicles (SUVs), the cacophony of billboards and donor logos in the streets, and the erection of expat compounds, drawing new boundaries of gated communities. The profound effects of this humanitarian sector on the local urban economy, service delivery and the dynamics of urbanization have been described in detail by Büscher and Vlassenroot (2010). Thus the humanitarian and development actions and interventions have shaped and reshaped the city in complex ways. A variegated landscape of international NGOs, civil-society actors and donor agencies is engaged in urban practices of relief, development, service delivery, aid and peace, and this has formed a new institutional landscape. As such, Goma can be perceived as an illustration of 'humanitarian urbanism': 'A novel urban condition that results from the expansion of the spatial, chronological and ideological scopes of humanitarian interventions' (Potvin 2013: 3).

Goma, the urban 'peaceland': A city where peace takes place

While Goma is perceived as the centre of local war economies, it is at the same time portrayed as the urban centre of a peacebuilding economy. Both dynamics make of Goma a city that, by its strategic position in local and global interventions of war and peace, is strongly connected to the global.

'Peacekeeping', 'peacebuilding', 'peace enforcement' – translated in actions, initiatives, programmes, training, projects, operations – are expressions of Goma's current 'global connectedness', as the central node of the international community's engagement in the region.

Over the last ten years, Goma has especially been portrayed as a symbol of failure with regard to peacebuilding (Autesserre 2007, 2010; Tull 2009). The capture of the city by the M23 rebels in November 2012 and the failure of the international peacekeeping force, MONUSCO, to protect the urban inhabitants was another painful example of this. The remaining problematic reputation of FARDC despite the huge investments in capacity-building and training is yet another. Failed demobilization campaigns affect the urban security situation in Goma as groups of *démob* (demobilized soldiers) aimlessly hang around in town and survive by begging and harassing small traders (Lange and Kaminuka 2010).

In that sense, Goma can be perceived as a city where peace fails to take place. Many of the donor agencies and aid, development and humanitarian organizations with headquarters in Goma have over the years reoriented (reinvented) themselves towards this peacebuilding narrative and have become part of it.

Recent in-depth ethnographic analyses of the 'peacebuilding apparatus' in Goma have produced interesting presentations of the city as
an urban 'peaceland', or a complex of international peace institutions,
individuals, policies and practices (Autesserre 2014; Jennings 2014).
This research has beautifully demonstrated how this peacebuilding
economy and the actors involved in it constantly interact with, and
inevitably shape, the local reality in which they operate (Jennins
2014). 'Peaceland', echoing the concept of 'aidland' developed by Mosse
(2011) and Fechter and Hindmann (2011), is in the first place perceived as an enclosed expat world, a bubble, a separate world with
its own time, space, economy, culture, rituals, behaviours and beliefs
(Autesserre 2014). Large numbers of the peacekeepers, peacebuilders,
humanitarians, development workers and others are expats. However,
the 'peacebuilding apparatus', of course, does not only involve expatriates; international attention and money invested in projects related
to peacemaking in Goma has produced a local NGO and association
scene specifically engaged in peace issues. The *Amani* label (Kiswahili
for 'peace') is omnipresent at local NGO and ASBL (association sans but
lucratif/association without lucrative purpose) desks, training and programme banners, and it is, of course, echoed in local football teams,
music bands and so on. The locally organized (and partly internationally
funded) Amani Festival (since February 2014, taking place every year),
gathering the urban youth and those from across the border around
music, dance and performance, was a colourful celebration of Goma as
the urban symbol of peace (All Africa 2014).

Furthermore, the local embedding of 'peaceland' is very outspoken
when considering Goma's strongly developed 'peacebuilding economy';
cash flows linked to international peace investments entering town give
rise to particular local urban markets, responding to specific demands
within this 'peace enterprise'. Goma's inhabitants, both small-scale
informal entrepreneurs and influential economic elites, invest their
money in this market (e.g. by opening shops with specialized expat-
demand-merchandise, by organizing Amani activities and by building
houses to rent to expat staff).

The ultimate embodiment of the urban 'peaceland' is the international peacekeeping force, MONUSCO. Present in Goma since 2000,
it gradually developed into the largest peacekeeping mission globally
with 19,000 troops and more than 3,000 (UN Peacekeeping 2015) personnel, of which large parts are based in Goma. The white tanks and
blue helmets have become an integral part of the city's standard image.
MONUSCO's image, credibility and legitimacy in Goma, and in the

region in general, have been the subject of much critical discussion (Autesserre 2010; Koddenbrock 2012; Quintiliani 2014), but studies on the interaction between this peacekeeping force and the local urban environment have been very limited (Autesserre 2014; Boas 2013). However, the visible presence and impact of the social, economic and spatial architecture created by peacekeeping and peacekeepers in Goma cannot be denied (Boas 2013; Büscher and Vlassenroot 2010).

Despite the fact that much of the peacebuilding activities and programmes (just like their humanitarian counterparts) have their focus outside the urban centre, the heart of 'peaceland' is the city centre itself, where one finds most of their work and residing infrastructure. The impact on the city's socioeconomic, political and spatial profile is generated as much by the peace actors' presence as by their actions. The separate world or 'bubble' in which many 'peaceland inhabitants' live can be observed from the urban landscape in the form of walled UN compounds and luxurious expat infrastructure, with strong investments in particular urban neighbourhoods (namely the districts Volcans and Himbi, where much of the humanitarian and peacebuilding offices and staff are located). So the social disconnection between expat 'peacelanders' and the surrounding urban environment is strongly reflected through these enclosed spaces (Duffield 2010; Smirl 2008). Expat staff in Goma often have strict rules regarding their interaction with the 'local context' and are simply not allowed to access many districts of Goma town. Another typical marker of the humanitarian, aid and peacebuilding industry is the booming 'entertainment infrastructure' in Goma. Luxurious hotels, bars, restaurants and nightclubs have mushroomed in number over the past ten years, with many again almost exclusively oriented towards an international clientele. The international expat community perceives the city as a place to rest, relax and sleep (Büscher and Vlassenroot 2010).

Spatial agency by peacekeeping and peacebuilding actors happens most of the time in an indirect and unintended way – for example, by the allocation of large strips of public space by the Congolese state to the UN for the development of compounds and military bases. When these actors actively engage themselves in 'spatial' actions such as the construction of infrastructure or schools, these actions are often not urban based. The production of urban space by these actors thus occurs in a far more indirect but nonetheless influential way. Yet peacebuilding actions and practices have been strongly integrated into Goma's current cityscape.

Local agency and spatial production: Goma, city of prosperity and touristic nostalgia

Hitherto I have demonstrated how Goma embodies both dynamics of violent conflict and peace, expressed in its political, socioeconomic and spatial urban profile. Let me now focus on local spatial agency that transcends this binary image of Goma as a city where conflict and peace take place. If one asks Goma's inhabitants about their city, they often will stress especially all other things taking place in Goma – other than rebellion and peacebuilding. When circulating through town with urban youngsters, they will show you what they perceive to be the main 'points of interest' in town. Besides the UN base, you will probably learn about the new ultramodern supermarket, the new football stadium, the new multistorey apartment building, the new fancy hotel, the new rehabilitated road, the new statue and the just opened new international bank.

Young 'Goméens', in their discourse on the current development of their town, will stress issues such as 'opportunity', 'development' and 'business'. This construction of an urban image or identity from within is also reflected in an active involvement in urban place-making.

Goma as an extremely fast-growing and developing city displays a rapidly changing urban landscape. Spatial developments there need to be framed in the particular context of an extremely weak state and a profound informality on all levels of urbanization. In academic debates on both state failure and informal urbanism in Africa, the DRC is an often-cited example (Lemarchand 2001; Mathews and Solomon 2001; Nsokimieno Eric et al. 2014; Reno 2006; Trefon 2004). The urbanization process of Goma does take place in a very informal way, but not completely outside state control. The Congolese state, represented by the local administration, the cadastre office and the ministry of land, among others, is involved in issues such as urban planning, but its actual implication is weak. Examples are the uncontrolled informal urbanization with large concentrations of illegal constructions and the strong involvement of non-state actors in urban service delivery (e.g. in terms of education, security or urban infrastructure).

Processes of state failure (reinforced by a protracted situation of civil war) have facilitated the development of a number of non-state actors that have replaced the state in a number of crucial responsibilities with regard to urban governance. These actors became increasingly involved in the organization of public space and in local processes of political and socioeconomic regulation (Vlassenroot and Raeymaekers 2008). In the particular context of Goma, local urban 'big men' and the international 'humanitarian/peacekeeping' organizations are the two most important non-state actors because they engage in different domains of urban governance and often act with more authority and legitimacy than the Congolese state.

With regard to spatial agency, while the active spatial interventions of the international humanitarian and peacebuilding actors are rather limited, this is not the case for the other influential non-state actor: the local urban elite or 'big men'. When investigating material spatial investments in Goma, the role of the state is much more marginal than that of businessmen, who are strongly engaged in urban economic and material investments. These urban 'big men' of Goma are a small group of influential businessmen of different ethnic and social backgrounds who are in control of today's main urban economic markets. They operate in strategic alliances with political and military elites, are often rich in both urban and rural land titles and real estate, and they reinvest their money in lucrative urban markets (e.g. natural resources and the importation of manufactured goods or cars). These big men increasingly engage in urban planning and the active shaping of the physical urban landscape. Participating in urban infrastructural or development projects, for example, is part of these big men's power and authority strategies in Goma.

To be respected (and also protected) in this complex urban environment, these elites need to constantly invest in social capital among the urban population. They do this by presenting themselves as initiators of development and wealth, and donating to projects investing in, for example, urban infrastructure that symbolize prosperity and development. By investing in the city's architecture, asphalting streets, rehabilitating neighbourhoods, erecting statues and so forth, elites display their authority by signing the landscape and imposing their name. For example, in 2010, several statues were erected at the different roundabouts in town as 'donations' of Goma's *grands barons*. The surrounding districts now carry their names. These investments in urbanization are presented as an expression of their 'devotion' to the city and its inhabitants. Through the transformation of the urban space, these actors materialize their belief in, and contribution to, Goma as a city of progress, development, hope and prosperity. Goméens greatly appreciate these kinds of initiative and perceive them as a sign of goodwill and the big men's commitment to the city's development. When these elites, for example, invest their money in yet another luxurious hotel, inhabitants see this as their commitment to investing locally, so that not all the advantages of wealth (even if this wealth is created by their indirect or direct involvement in the dynamics of war) are transferred abroad.

The urban landscape that has been produced by these elite's spatial interventions is one that tends to symbolize progress, prosperity and development. The idea of Goma as a booming city of opportunity, despite its location in a crisis region, is another image of the city that is often emphasized by the urban younger generations: an attractive place where one is able to find a good living; a fast-developing, globally connected city of possibility and promises. Goma embodies the paradoxical reality of the productive power of violent conflict (Büscher and Vlassenroot 2013). The huge gold-coloured monument of the *tshukudu*[3] represents the 'brave working spirit' of Goma's inhabitants. *Tshukudeurs*, who fill the streets of the popular informal economy districts of town with their wooden bikes piled up with loads of vegetables, charcoal or palm oil, have become a symbol of Goma's spirit of creative coping strategies but also of the city's fertility and opportunities. They are seen, for example, as the thriving force behind local commercial opportunities. Further, multilevel stores and beautiful modern houses that are continuously being constructed are also referred to as proof of local entrepreneurship, a growing class of rich businessman and so forth. It is indeed remarkable how fast these modern villas pop up in several of Goma's neighbourhoods.

Another urban image that is actively reproduced by spatial interventions is that of Goma's historical 'grandeur' as an attractive tourist city. During colonial times, Goma and its Rwandan 'twin city', Gisenyi (located on the other side of the border), had developed into important tourist attractions where Belgians from all over the colonies came to spend a relaxing holiday. A 1958 Belgian travel guide for the Belgian Congo and Rwanda-Urundi states:

> Goma and Kisenyi complement each other beautifully. They are destined to become a very important tourist centre and a valued holiday and retreat resort. Goma is built on the lava . . . on the other hand, Kisenyi is endowed with a beautiful beach in a setting of greenery and flowers . . . Not one place in Central Africa can compete in the tourism sector with Goma and Kisenyi; they are rich in natural beauty: a marvelous lake, a lovely beach, one of the most remarkable volcanic regions in the world, . . . near National Parks, with an abundant variety of fauna, ideal climate and varied entertainment.
>
> (Infor Congo 1958: 645, my translation)

From Goma, tourists could visit the islands on Lake Kivu, the Parc National Albert and the volcanoes. At the end of the colonial period, when the Belgian presence was at its greatest, important investments were made in touristic infrastructure, which were further expanded in the 1970s and 1980s during the period of President Mobutu's reign. The impressive Hotel des Grands Lacs reminds us of these colonial touristic high days. The building is still present and used as a hotel, and in some rooms one can even find some of the original 'colonial' bathtubs. This touristic importance of Goma is still strongly alive in local urban imaginations. While strolling around in town and admiring the stunning view over Lake Kivu and the beautiful surrounding landscapes in the city's background (the volcanoes, the Masisi hills), it is easy to imagine the picturesque tourist attraction Goma once used to be. Spatial investments 'from within' tend to reproduce this image, through the construction of tourist infrastructure, but also by more direct markers in the urban landscape, such as the recently refurbished roundabout in the city centre representing an idyllic picture of Nyragongo Volcano surrounded by its natural parks, lions and elephants. Such examples are a spatial celebration of Goma's pre-war image as a zone of natural beauty and peace.

It is remarkable how Goméens continue to identify their city with these remnants of past glory, invariably presenting their city in the first place as *une ville touristique*. For a Congolese, 'Goma' still connotes tourism. In the capital city, Kinshasa, Goma is perceived as their remote province, known for its rebels, its perfect climate and its touristic attractions. References to Goma's capacity as a tourist town are to be found not only in the hotel scene and architectural symbols but also in, for example, the omnipresence of the officials of the ministry of tourism. The provincial and municipal branch of this ministry is one of the most overrepresented administrative state agencies in Goma (and its surroundings). In fact, outsiders require a permission from the *office de tourisme* to move freely through the city, to enter buildings or infrastructures, to buy particular goods and to talk to particular people.

However, this spatial reproduction of a tourist city is not only the outcome of spatial agency from 'within'. The large-scale presence of international donor, humanitarian and peacekeeping agencies has obviously given new impetus to the existing tourist infrastructure; as I mentioned, the spectacular increase in hotels, bars and restaurants is directly linked to the urban humanitarian and peacekeeping economy. In Goma it is in fact hard to find 'ordinary' tourists (frightened by the persisting insecurity), yet the development of an extensive infrastructure of 'humanitarian tourism' has largely replaced these tourist by 'peace-landers' or humanitarian and development aid staff ('attracted' in a way by the persisting insecurity). They perceive the city as a place to rest, relax and sleep. Expats used to call Goma *la ville phare*, referring to its varied nightlife. Reconnecting to the city's pre-war image of a tourist attraction, the remaining tour operators today offer boat trips or waterskiing on Kivu Lake, or picnics on the small islands, to Goma's permanent expat staff.

Concluding remarks

Interestingly, this demonstrates how the image of Goma as a touristic city of opportunities is in fact an indirect effect of both dynamics of war economy and peacebuilding economy. The 'alternative' urban image that is produced and reproduced socioeconomically and spatially from within actively counters the perception of Goma as a city of violent conflict and post-conflict reconstruction. Yet it is in itself the outcome of these same dynamics that I analytically framed as part of 'no-war-no-peace' societies. The 'humanitarian tourism' infrastructure as well as the local booming economy are clearly outcomes of an urbanization in which 'conflict and peace take place', yet the city's urban inhabitants symbolize its hope, its business opportunities and its progress.

This chapter has presented an alternative reading of a city that in academic and donor literature often serves as a case study for local conflict and peace dynamics in the complex Kivu region in eastern DRC. By perceiving Goma as an urban 'no-war-no-peace' society and particularly focusing on the intended and unintended production of space by different urban actors involved in the dynamics of war and peace and beyond, I have taken the urban landscape as a lens through which to study local socioeconomic and political urban transformations. Investigating local conflict and peace dynamics through this particular lens can teach us more about the strategies of local and global actors involved in war-and-peace practices on a very local scale. This approach demonstrates how different actors, by engaging in very different spatial interventions, produce different images of 'their' city, going beyond the externally constructed binary interpretation of Goma as a city of rebellion or a city of failed peacebuilding initiatives.

To study the spatial impact of local war-and-peace dynamics, urban 'no-war-no-peace' zones remain fascinating yet complicated research sites, requiring methodological flexibilities and interdisciplinary approaches. Further research on the material translation of the particular forms of urbanism that are produced by the protracted dynamics of civil war and peacebuilding initiatives forms an indispensable contribution to the broader debates among political and human geographers, political anthropologists and conflict researchers about the relation between space, conflict and peace.

Notes

1. By 'expat' I refer to this category of international staff employed by international humanitarian organizations, the UN and other donor agencies, NGOs and peace-keeping forces. They are employed under so-called 'expat contracts', they refer to themselves with this terminology and are locally referred to as 'expats'.
2. A Rwandan-backed rebel movement that was in control of large parts of eastern DRC between 1999 and 2004.
3. A wooden scooter to transport heavy loads in the absence of motorized transport.

References

All Africa (2014) *Congo-Kinshasa: The Amani Festival Shows Off Goma's Fun Side*, http://allafrica.com/stories/201403040613.html (accessed 8 July 2015).

Autesserre S (2007) D.R. Congo: Explaining Peace Building Failures, 2003–2006. *Review of African Political Economy* 34(113): 423–442.

Autesserre S (2008) The Trouble with Congo: How Local Disputes Fuel Regional Conflict. *Foreign Affairs* 87(3): 94–110.

Autesserre S (2010) *The Trouble with the Congo: Local Violence and the Failure of International Peacebuilding*. New York: Cambridge University Press.

Autesserre S (2014) *Peaceland: Conflict Resolution and the Everyday Politics of International Intervention*. New York: Cambridge University Press.

Beall J Goodfellow T and Rodgers D (2011) *Cities, Conflict and State Fragility. Working Paper No. 85*. London: London School of Economics and Political Science.

Berwouts K (2012) *Goma falls to the M23: a tale of war, rebellion and dreadful peace agreements*, http://africanarguments.org/2012/11/21/goma-falls-to-the-m23-a -tale-of-war-rebellion-and-dreadful-peace-agreements—by-kris-berwouts/ (accessed 17 November 2015).

Boas M (2013) A Tale of Two Cities: The Peacekeeping Economy of Goma and Monrovia. In: *5th European Conference on African Studies*, Lisbon, Portugal, 26–29 June 2013.

Boas M and Dunn K (2014) Peeling the Onion: Autochthony in North Kivu, DRC. *Peacebuilding* 2(2): 141–156.

Büscher K (2011) *Conflict, State Failure and Urban Transformation in the Eastern Congolese Borderland: The Case of Goma*. PhD Dissertation, Ghent University, Belgium.

Büscher K (2012) Urban Governance Beyond the State: Practices of Informal Urban Regulation in the City of Goma, Eastern DR Congo. *Urban Forum* 23(4): 483–99.

Büscher K and Mathys G (2013) Navigating the Urban 'In-Between Space': Local Livelihood and Identity Strategies in Exploiting the Goma/Gisenyi Border. In: Korf B and Raeymaekers T (eds) *Violence on the Margins: States, Conflict, and Borderlands*. New York: Palgrave Macmillan, pp. 119–142.

Büscher K and Vlassenroot K (2010) Humanitarian Presence and Urban Development: New Opportunities and Contrasts in Goma, DRC. *Disasters* 34(2): 256–273.

Cramer C (2006) *Civil War Is Not a Stupid Thing: Accounting for Violence in Developing Countries*. London: Hurst & Co.

De Saint Moulin L (2010) Villes et Organisation de l'Espace au Congo (RDC). *Cahiers Africains, 77*. Tervuren: Musée royal de l'Afrique centrale.

Dijkzeul D (2008) Towards a Framework for the Study of 'No War, No Peace' Societies. *Swiss Peace Foundation Working Paper* 2/2008, Bern: Swiss Peace Foundation.

Doevenspeck M (2011) Constructing the Border from below: Narratives from the Congolese-Rwandan State Boundary. *Political Geography* 30(3): 129–142.

Doevenspeck M (2013) An Impossible Site? Understanding Risk and Its Geographies in Goma, Democratic Republic of Congo. In: Müller-Mahn D (ed.) *The Spatial Dimension of Risk: How Geography Shapes the Emergence of Riskscapes*. Abingdon: Routledge, pp. 137–153.

Duffield M (2001) *Global Governance and the New Wars: The Merging of Development and Security*. London: Zed Books.

Duffield M (2010) Risk-Management and the Fortified Aid Compound: Everyday Life in Post-Interventionary Society. *Journal of Intervention and Statebuilding* 4(5): 453–474.

Dunn KC (2001) Identity, Space and the Political Economy of Conflict in Central Africa. *Geopolitics* 6(2): 51–78.

Eichstaedt P (2011) *Consuming the Congo: War and Conflict Minerals in the World's Deadliest Place*. Chicago: Lawrence Hill Books.

Fechter AM and Hindman H (eds) (2011) *Inside the Everyday Lives of Development Workers: The Challenges and Futures of Aidland*. Sterling: Kumarian Press.

Infor Congo (1958) *Belgisch-Congo en Ruanda-Urundi. Reisgids*. Brussel: Infor Congo.

IRIN (Integrated Regional Information Networks) Africa (2012) DRC: *Growing Humanitarian Needs in Goma*, http://www.irinnews.org/report/96913/drc-growing-humanitarian-needs-in-goma (accessed 8 July 2015).

IRIN Africa (2013) Goma Running out of Space for DRC's Displaced, http://www.irinnews.org/report/98624/goma-running-out-of-space-for-drc-s-displaced (accessed 8 July 2015).

Jackson S (2002) Making a Killing: Criminality & Coping in the Kivu War Economy. *Review of African Political Economy* 29(93–94): 516–536.

Jackson S (2003) Fortunes of War: The Coltan Trade in the Kivus. In: Collinson S (ed.) *Power, Livelihoods and Conflict: Case Studies in Political Economy Analysis for Humanitarian Action:* HPG Report 13, February. Available at: http://www.odi.org/sites/odi.org.uk/files/odi-assets/publications-opinion-files/289.pdf (accessed 2 July 2015), pp. 22–36.

Jackson S (2006a) Borderlands and the Transformation of War Economies: Lessons from the DR Congo. *Conflict, Security & Development* 6(3): 425–447.

Jackson S (2006b) Sons of Which Soil? The Language and Politics of Autochtony in Eastern D.R. Congo. *African Studies Review* 49(2): 95–123.

Jennings K (2014) Service, Sex and Security: Gendered Peacekeeping Economies in Liberia and the Democratic Republic of the Congo. *Security Dialogue* 45(4): 313–330.

Keen D (1998) The Economic Functions of Violence in Civil Wars. *The Adelphi Papers* 38(320): 1–88.

Koddenbrock K (2012) Recipes for Intervention: Western Policy Papers Imagine the Congo. *International Peacekeeping* 19(5): 549–564.

Lamarque H (2014) Fuelling the Borderland: Power and Petrol in Goma and Gisenyi. *Articulo - Journal of Urban Research* 10. Available at: https://articulo.revues.org/2540 (accessed 2 July 2015).

Lambach D (2007) Oligopolies of Violence in Post-Conflict Societies. *GIGA Working Paper* No. 62. Hamburg: GIGA. Available at: http://www.giga-hamburg.de/en/system/files/publications/wp62_lambach.pdf (accessed 2 July 2015).

Lange M and Kimanuka C (2010) *The Crossing. Small Scale Trade and Improving Cross-Border Relations between Goma (DR Congo) and Gisenyi (Rwanda)*. London: International Alert.

Larmer M, Laudati A and Clark JF (2013): Neither War Nor Peace in the Democratic Republic of Congo (DRC): Profiting and Coping Amid Violence and Disorder. *Review of African Political Economy* 40(135): 1–12.

Le Billon P (2001) The Political Ecology of War: Natural Resources and Armed Conflicts. *Political Geography* 20(5): 561–584.

Le Billon P (2013) *Fuelling War: Natural Resources and Armed Conflict. Adelphi Paper* 373. Abingdon: Routledge for IISS.

Lemarchand R (2001) The Democratic Republic of Congo: From Collapse to Potential Reconstruction. *Occasional Paper.* University of Copenhagen, Centre for African Studies. Available at: http://teol.ku.dk/cas/publications/publications/occ._papers/lemarchand2001.pdf (accessed 29 June 2015).

Mac Ginty R (2006) *No War, No Peace: The Rejuvenation of Stalled Peace Processes and Peace Accords*. Basingstoke: Palgrave Macmillan.

Mampilly ZC (2011) *Rebel Rulers: Insurgent Governance and Civilian Life during War*. Ithaca: Cornell University Press.

Mathews S and Solomon H (2001) The Challenges of the State Collapse in Africa: The Case of the Democratic Republic of Congo. *Africa Insight* 31(3): 1–12.

Mosse D (2011) *Adventures in Aidland: The Anthropology of Professionals in International Development*. Oxford: Berghahn Books.

Mpisi J (2008) *Kivu, RDC: La paix à tout prix! La Conférence de Goma (6–23 janvier 2008)*. Paris: L'Harmattan.

Nsokimieno Eric MM, Bailey E, Martin, MT, Abraham C, Li J, and Liqin Z (2014) Ongoing Informal Settlements in Democratic Republic of Congo: Implementing New Urban Policy for Creating Sustainable Neighborhoods. *Journal for Sustainable Development* 7(5): 254–265.

Oldenburg S (2010) Under Familiar Fire: Making Decisions during the 'Kivu Crisis' 2008 in Goma, DR Congo. *Africa Spectrum* 45(2): 61–80.

Pole Institute (2007) '*Nord Kivu: le triomphe de la politique du pire?*' (16/10/07), http://www.pole-institute.org/site%20web/echos/echo64.htm (accessed 7 July 2015)

Potvin M (2013) Humanitarian Urbanism under a Neoliberal Regime: Lessons From Kabul (2001–2011). In: *International RC21 Conference 2013*, Berlin, Germany, 29–31 August 2013.

Quintiliani P (2014) What It Takes to Bring Peace to the Eastern DRC. *Conflict Trends* 2014 (1): 43–9. Available at: http://www.accord.org.za/images/downloads/ct/ACCORD-Conflict-Trends-2014-1_p43_Eastern_DRC.pdf (accessed 2 July 2015).

Raeymaekers T (2010) Protection for Sale? War and the Transformation of Regulation on the Congo-Ugandan Border. *Development and Change* 41(4): 563–587.

Reno W (2006) Congo: From State Collapse to 'Absolutism' to State Failure. *Third World Quarterly* 27(1): 43–56.

Richards P (ed.) (2005) *No Peace No War: An Anthropology of Contemporary Armed Conflicts*. Oxford: James Currey.

Robinson J (2006) Inventions and Interventions: Transforming Cities – An Introduction. *Urban Studies* 43(2): 251–258.

Rohwerder B (2013) Democratic Republic of the Congo: Internally Displaced Persons and Refugees' Relations with Host Communities. GSDRC Helpdesk Research Report. Available at: http://www.gsdrc.org/docs/open/HDQ1032.pdf (accessed 29 June 2015).

Smirl L (2008) Building the Other, Constructing Ourselves: Spatial Dimensions of International Humanitarian Response. *International Political Sociology* 2(3): 236–253.

Swart G (2011) No War, No Peace in the Volatile East. *Peace Review: A Journal of Social Justice* 23(2): 144–153.

Themnér A (2011) *Violence in Post-Conflict Societies: Remarginalization, Remobilizers, and Relationships*. Abingdon: Routledge.

Trefon T (ed.) (2004) *Reinventing Order in the Congo. How People Respond to State Failure in Kinshasa*. London: Zed Books.

Tull D (2005) *The Reconfiguration of Political Order in Africa. A case Study of North Kivu (D.R. Congo)*. Hamburg African Studies 13. Hamburg: Institut für Afrika-Kunde.

Tull D (2009) Peacekeeping in the Democratic Republic of Congo: Waging Peace and Fighting War. *International Peacekeeping* 16(2): 215–230.

UN Peacekeeping (2015) *Peacekeeping Fact Sheet*, http://www.un.org/en/peacekeeping/resources/statistics/factsheet.shtml (accessed 8 July 2015).

Van Acker F (2005) Where Did All the Land Go? Enclosure & Social Struggle in Kivu (D.R. Congo). *Review of African Political Economy* 32(103): 79–98.

Verhoeve A (2004) Conflict and the Urban Space: The Socio-economic Impact of Conflict on the City of Goma. In: K. Vlassenroot and Tim Raeymaekers (eds) *Conflict and Social Transformation in Eastern Congo*. Gent: Academia Press, pp. 103–122.

Verweijen J (2013) Military Business and the Business of the Military in the Kivu's. *Review of African Political Economy* 40(135): 67–82.

Vlassenroot K and Büscher K (2013) Borderlands, Identity and Urban Development: The Case of Goma (Democratic Republic of the Congo). *Urban Studies* 50(15): 3168–3184.

Vlassenroot K and Raeymaekers T (2008) New Political Order in the DR Congo? The Transformation of Regulation. *Afrika Focus* 21(2): 39–52.

Yanacopulos H and Hanlon J (eds) (2006) *Civil War, Civil Peace*. Oxford: James Currey.

5

'The Camp', 'The Street', 'The Hotel' and 'The Karaoke Bar/Brothel' – The Gendered, Racialized Spaces of a City in Crisis: Dili, 2006–2008

Henri Myrttinen

Introduction

Space, its perception and the relative attributes linked to it, be it danger, safety, comfort or alienation, is highly personalized and time-bound, and informed by gender, class, age and what, for lack of a better term, I will call here 'racialized dynamics'.[1] My thoughts here are an attempt to make sense of my repeated visits to Dili, the capital of East Timor, focusing on the 'leaden years' of 2006–2008.[2] Those years were marked by an extremely visible presence of over 100,000 IDPs in makeshift camps across the city and its environs; a militarized international peacekeeping[3] and national security force presence; pervasive fear, especially between April 2006 and December 2006; street-fighting emanating from gangs, martial arts groups (MAGs) and ritual arts groups (RAGs); and a large international presence in the form of the UN Integrated Mission in Timor-Leste, international agencies such as the International Organization for Migration and a plethora of international NGOs (INGOs) with thousands of expat staff, locally referred to as *malae* in Tetum.[4]

The crisis, often referred to locally as the *krize* (crisis) or *situasaun* (situation), was triggered by protests in April 2006 by a group of around 600 soldiers who became known as the 'petitioners', as they had petitioned the government to look into accusations of their alleged discrimination based on their regional provenance from the west of the country. Their protests escalated in late May and became a catalyst for violent attempts to settle a range of issues and grievances: between different factions within the armed forces and the police; between the armed forces and the police; between different patronage networks among the political elite; between neighbours bearing long-standing grudges; between perceived winners and losers of the post-independence settlement; between rival gangs, MAGs and RAGs; between 'Easterners' and 'Westerners'; between supporters of different political parties; between young East Timorese street-fighting men and international peacekeepers, to name a few fissures. Due in part to an external intervention by peacekeeping forces, the violence which peaked late May/early June 2006 slowly subsided, but street-fighting and insecurity continued to de-stabilise Dili until early mid-2008.

The crisis itself, its microdynamics and aftermath have been documented elsewhere (for a detailed description of events, see e.g. UN 2006), but what has been missing to date, to my knowledge, is a gendered spatial analysis that examines the way different groups sharing the same space of the city lead tangentially connected lives geographically, even though their lives are intimately intertwined. While this sounds somewhat abstract, it is often very much everyday life in similar settings where, as in the case of Dili, the geographical distance between the well over 1,000 IDPs sleeping under tarps and makeshift shelters for a second year running and their purported, highly salaried helpers sipping imported coffee and beer inside the air-conditioned Hotel Timor was less than 15 m, but they might as well have been on different planets. The reality of the people visible from the lobby café was mostly only discussed in the abstract. In addition to 'The (IDP) Camp' and 'The Hotel' as ideal types, there were also numerous other spaces. I will consider two of them – 'The Street', where much of the fighting between various gangs, MAGs and RAGs, and the security forces occurred, and 'The Karaoke Bar/Brothel', where some of the non-East Timorese yet 'non-international' (i.e. non-OECD (Organisation for Economic Co-operation and Development) nationals, not UN or INGO affiliated) male foreigners (often referred to as 'third-country nationals')[5] spend their time and earnings and which doubled either directly as venues for sexwork or as places for soliciting such activities.

This chapter is not based on dedicated research on this issue *per se* but rather a reflection on the time I spent in these spaces while in East Timor. I was working for over a year for various NGOs, as a freelance journalist, as well as carrying out academic research on masculinities and violence. This chapter is an exploratory sketch, based retrospectively on my memories of a city which by now has in many ways changed greatly – the IDP camps are gone, the fear has mostly evaporated, the UN mission has left and the government, with Chinese help, is busily erecting flashy administrative buildings.

This *ex-post* reflection on the gendered, racialized and class-based defined access to various spaces builds on initial thoughts which I had during my numerous visits to and stays in Dili during the crisis years of 2006–2008, but it was only later that I considered them through a spatial lens. As this contribution is about the various social markers that mediate access (or lack of it) to different spaces, it is necessary for me to place myself and my own access and privilege firmly into the picture. It is also a sketch seen through my own gendered, class-based, racialized eyes – those of a (back then) early-30s, North European, male middle-class researcher/NGO worker, with all the attendant privileges, limitations and prejudices, and reflected upon through the refracted lens of memory half a decade later. As someone with NGO/media accreditation, I had access to official buildings (e.g. various ministries, embassies and the UN headquarters in the Obrigado Barracks) and officials; as a Westerner, I had the social and financial capital necessary to gain access to restaurants, cafés, bars and expat parties; doing research on gangs and MAGs gave me at least limited access to these groups; and being a youngish man allowed for conversations with men and women as well as access to spaces which might have been very different were I a female researcher, older or in a more official position.[6]

Apart from the four 'ideal-type' spaces I cover here, there were, of course, a plethora of other spaces in Dili – for example, the barracks of the international and national security forces; the quieter neighbourhoods of the city, some richer, some poorer; the curry houses catering to South Asian staff of the UN mission and the East Asian eateries catering to the East Asians (including diasporic Chinese from Malaysia, Indonesia and, of course, East Timor); and the markets, on which some of the conflicts centred, or the slightly more multicultural meeting places, such as the arts centre of Arte Moris and, to an extent, the Motion nightclub, where younger expats would mingle with those East Timorese (mainly young and male) who had the necessary social and cultural capital to join in. To some of these I had access; to others I did not.

Gender, class, 'race' and space

At the risk of stating the obvious, my approach to the social identity markers of class/rank, gender, age, ethnic background and nationality is an intersectional one. These different factors are interrelated and code-fine many of the possibilities open to us, but also possible restrictions and vulnerabilities, as social actors in a given situation. While these factors set the framework within which we move and act, they are not fixed and immutable, nor do they wholly predetermine our actions.

In spite of a body of insightful, critical, feminist-Marxist analysis of gender, class and space that goes back at least to Massey's (1994) work from the 1970s onwards, the gendered study of post-conflict interventions has seldom paid attention to spatiality, with some exceptions, such as Higate and Henry (2009) and, significantly, Smirl (2009, 2015). Yet, as these authors point out, access to space and the naming of space in post-conflict and post-disaster interventions is highly gendered, racialized and dependent on one's socioeconomic status.

Some, such as expat men and to a slightly lesser degree women (depending on the cultural circumstances and safety concerns), are often free to roam more or less wherever they want, moving about in air-conditioned SUVs, in taxis or on foot if they are on their morning run. Depending on their status, they may also have access to helicopter travel as well as overseas rest and recreation. Their accommodation is often in gated, guarded compounds to which 'locals' have no access, unless they are guards, cleaners or lovers. At the other end of the spectrum of mobility were, for example, the East Timorese teenagers I interviewed in 2007 who had never even been to the seashore, which was less than 1 km away, citing security fears and lack of transport.

Gendered, racialized and class-related power differences aside, land-scapes and spaces are inscribed with various layers of meaning, invisible to some, highly present to others. These may be memories of violence and trauma, of contestations of meaning, of abodes of ancestral spirits, arenas of nation- and state-building, or sources of power in and of them-selves (Bovensiepen 2009; Kingsbury 2009; Stead 2012, 2014; and for a regional but non-East Timorese view, see Ballard 2002). These would be mostly visible to the East Timorese, less so to the foreign presence. The different geographies were also visible in the way people navigated the city: whereas it can be safely assumed that the majority of expats would know where the Castaways Bar was, few East Timorese, including taxi drivers, would; conversely, the Brimob compound (still named after the Indonesian Brigade Mobil paramilitary police force housed there until 1999) would be known to locals but not necessarily to *malae*. The new official names of streets, such as Avenida dos Martires da Patria or Rua do Portugal, would be mostly unknown to locals and foreigners alike.

What I will attempt here is to take the reader along for a small tour of the spaces as I experienced them. This 'guided tour' can in no way be a comprehensive history of Dili or even of a particular moment in time, but I hope to use it to raise issues around the gendered, class-based and racialized spatiality and materiality that define the lives of people in conflict- and disaster-affected societies. These differences are clear for all to see but are seldom theorized. Let alone is their impact on the way interventions are imagined, conceptualized and implemented taken into account by the international interveners.[7]

Introduction to Dili

Ever since the capital was moved to Dili by the Portuguese colonial administration in 1769, the city, built on a thin stretch of flat land between the sea and the mountains, has functioned as the territory's administrative and economic centre under the subsequent Portuguese, Japanese, Indonesian, UN and Timorese administrations. The city is divided into several *bairos*, which tend to become increasingly village-like with increasing distance from the colonial-age buildings lining the city centre's seafront. At the time of the crisis, the city had a rapidly growing population of around 212,000 inhabitants (Direcção Nacional de Estatística 2009: 12).

While the city centre has seen an increasing amount of urban development, such as the refurbishing of colonial-era administrative buildings and the construction of new ones, much of the rest of the city tends to have a ramshackle appearance. Burnt-out shells of houses destroyed in the militia violence of 1999 and the communal violence of 2006 continue to line the streets. Basic services, such as running water, electricity and waste management, are often lacking. Food security is often ensured through backyard market gardens and the keeping of chickens and pigs, which often roam the streets.

Though it had been the Portuguese administrative capital for 158 years at the time, the urbanized area in Dili in 1927 was only a thin stretch perhaps 1.5 km long and at best several blocks of houses deep, set along the beachfront close to the Palacio de Gobierno. By 1950 the urbanized area covered perhaps 2.5 km × 2.0 km, and by 1964 the population of the city was a mere 7,000 inhabitants (Teague 1964: 117). The real period of urban and periurban expansion came between 1966 and 1990, and especially during the Indonesian occupation (1975–1999), when the area of the city doubled. During this time, traditional houses were replaced by concrete ones and more informal structures of corrugated iron and wood (Durand 2006: 114).

Much of the city was destroyed during the last wave of militia and Indonesian military violence in September 1999. At the time of my research, the damage from 1999 had not yet been fully repaired, especially in the outer sectors of the city. A new wave of house-burning and destruction has accompanied the *krize* which had gripped the nation in April–May 2006. While the aftermath of the 1999 militia violence brought an influx of new migrants to the city, the crisis that erupted in 2006 had in part the opposite effect, with some residents at least temporarily seeking refuge in quieter parts of the country. The various waves of displacement and conflict, from the forced displacement of the rural by Indonesian counterinsurgency campaigns in the late 1970 and early 1980s, to the forced displacements of 1999 and 2006, have heavily impacted the city and the occupation of land plots and houses, as has voluntary in-migration from rural areas (Stead 2014).

The main economic centres of the city are around the port, the administrative centre and what could be termed the 'central business district'. There is little in the way of industry in the city apart from small repair workshops, construction companies and several logistics centres. Dili is the main port in the country and has the only international airport. Many of the commercial interests there are in the hands of foreign nationals (including Australians, Chinese, Indonesians, Portuguese and Singaporeans) and, traditionally, also in the hands of members of the Chinese-Timorese minority. While the foreign *malae* and the better-off Timorese shop in the air-conditioned supermarkets and smaller shops, most inhabitants of Dili procure their goods from the numerous outdoor markets, which have become occasional flash-points for fighting between rival groups. Beyond the commercial and administrative centre, the city has a largely periurban character, with small gardens, orchards and banana groves mixing with simple housing structures. It is not uncommon for the residents to keep chickens and pigs in their backyards and to allow them to roam freely.

'The Camp'

A defining feature of life in Dili during the period covered here was the presence of the IDP camps around Dili, a constant reminder of the impacts of the crisis. At the height of the crisis, 30,000–50,000 were living in IDP camps in Dili, and in nearby Hera and Metinaro. Another 50,000–70,000 were living as IDPs in other parts of the country, though these were not necessarily all from Dili (OCHA 2007).[8] It was the periurban fringe and the IDP camps which were the focal points of much of the fighting between the various groups of gangs, MAGs and RAGs.

The crisis had thus produced a number of communities which were highly volatile, highly concentrated and highly politicized (in the general sense of the word). Some of the most volatile of these camps also happened to be located in some of the most vulnerable strategic points of the country: at the international airport, at the main port and at the heart of the government district, opposite the UN headquarters and straddling the country's main road next to the major army base in Metinaro.

During the period 2008–2009, which saw improved security, both the East Timorese government and international aid agencies were able to persuade several thousand IDPs to return home, by offering relocation packages, and, in some cases, exerting pressure (International Crisis Group 2009: 3). By late 2009, all IDP camps had been disbanded and the inhabitants had returned to their communities, but up to 45 per cent of returnees reported having experienced renewed conflict upon their return (Scambary 2009: 3).

Life in the camps depended in part on their size and location: those in the smaller camps, sometimes consisting of no more than two or three tents in the compounds of NGO offices or churches, depended on negotiating relationships with their hosts, whereas larger camps, such as Jardim, the airport camp and Metinaro, had thousands of inhabitants and their own internal social and political structures.[9] Tensions in the camps were often inflamed by party political differences and mutual distrust between the IDPs and the surrounding population, gang rivalries, personal conflicts and the daily stress of protracted camp life.

During the approximately two-and-a-half years of their existence, the larger IDP camps started developing into small communities of their own, with their own shops, markets, supply lines, social hierarchies and sometimes even *warungs* (street-side kitchen/food stall) selling food.[10] Most of these small-scale, above-board economic activities were carried out by women. Rumours also circulated of informal gambling dens and brothels being run in the camps, but there was no publicly available information to either confirm or refute these assertions.[11] Access to outsiders was limited and restricted, and for the most part the only foreigners entering the camps were aid workers and, in rare cases, media representatives or researchers.

While the majority of the camp population was from the poor urban classes which make up the majority of Dili's population, some middle-class people also resided in the camps, including government bureaucrats and local NGO staff. Those with steady jobs commuted to work from the camps, while others sought temporary jobs in the more informal sectors of the economy, access to which was negotiated through male-dominated networks of patronage, often linked to political parties as well as gangs, MAGs and RAGs. There was a relatively large degree of interaction between the major camps in Dili with IDPs, especially young men, shuttling between various camps. There were rumours going around at the time that some of these camps would in fact be turned into permanent settlements.

Political representation inside the camps and *vis-à-vis* the outside world was in the hands of men, with older men taking care of the more 'serious' negotiations behind the scenes and young men building up more visible pressure through public displays of militancy. As the camps became more established, they became incubators for political activity and often emanated a palpable sense of frustration and latent aggression at boiling point. The daily circumstances of the lives of the IDPs were a constant reminder of the fact that the crisis had not been resolved. In spite of the visible presence of the flags of the opposition party, FRETILIN (Frente Revolucionária de Timor-Leste Independente/Revolutionary Front for an Independent East Timor), at the time of the main field research, no one party was able to unite this frustrated, and thereby rudimentarily politicized, populace under its banner. The major, temporary unifying factor was often a perceived common outside 'enemy' that was seen as a threat to the IDPs in general, such as the international peacekeeping forces in cases where IDPs had been shot, or governmental and international agencies which were not seen as delivering aid fast enough. Although most of the violent protest and occasional attacks on UN Police (UNPOL) and the International Stabilisation Force (ISF) were carried out by young men in the MAGs and RAGs as they entered the sphere of 'The Street', they often had the backing and logistical support of older men and their mothers, sisters, wives and girlfriends.[12]

'The Street'

Much of the fighting between the rival gangs and other groups was mainly concentrated in the poorer parts of town, especially at the western end of the city, such as in Bairo Pite, Comoro, Delta, Pantai Kelapa and Fatuhada. Becora, at the south-eastern end of the city, was traditionally a tense area but remained mostly calm during most of the crisis years of 2006–2008. One of the most highly contested 'prizes' in the initial struggles between the various groups in April–June 2006 were the relatively modern real-estate development schemes (*perumahan*) erected for civil servants and their families during the Indonesian occupation in Delta and Bairo Pite (Scambary 2012). Although, as mentioned above, the young men involved in the fighting had the sometimes overt, sometimes tacit, sometimes reluctant support of women and older men, the space of 'The Street' was an almost exclusively male-dominated one as soon as fighting broke out. These battles would be between different groups of men publicly demonstrating their willingness to use violence, in the spectacle of either East Timorese gangs, MAGs and RAGs pitted against each other, or of the militarized masculinist police presence of the UNPOL units against these groups.

The economy of these urban or peri-/semiurban areas tends to be characterized by what is often referred to as the 'informal' sector, though many inhabitants do commute to work in other parts of the city for more 'formal' employment. However, unemployment tends to be high and demographically there is a bias towards the younger generations. The walls of the *bairos* of Dili are covered with slogans and graffiti, often referring to various gangs (thus acting as territorial markers) and, during the crisis, often with derogatory, anti-'Easterner' messages.

Many of the inhabitants of these periurban areas are first- or second-generation migrants to the city and the new migrants tend to move into areas where previous migrants from the same region settled, leading to a degree to a reproduction of village-level networks of support and loyalty (Scambary 2009, 2012). Increased migration to Dili from rural areas has been quoted by several East Timorese and international observers as a key factor in leading to unstable and unsettled (in a dual sense) neighbourhoods, which lack the social cohesion of older urban communities, thus becoming more susceptible to social phenomena such as gang violence (see e.g. Prüller 2008; Streicher 2008). Moxham (2008: 19) notes that 'those seeking opportunity in Dili were increasingly faced with acutely unequal and stagnant employment markets, a housing squeeze, unaffordable basic commodities and exposure to the volatile global rice market'.

One aspect of life in Dili for many East Timorese, which I found striking and which became even more pronounced during the political crisis, was the extreme locality and lack of mobility in most city-dwellers' lives. Especially for the socioeconomically less well-off citizens living in the *bairos*, much of their life tended to take place within a radius of perhaps a few hundred metres. This was due to both to imposed restrictions, such as a lack of modes of transportation and the prohibitive cost of taxis, and to self-imposed limitations engendered by the prevailing climate of diffuse fear ever since the outbreak of the crisis. Women's mobility, especially that of the East Timorese but also 'expats' and 'third-country nationals', was reduced by the crisis (Myrttinen 2008).

As Goddard (2005: 19–22) points out in the case of Papua New Guinea's capital city of Port Moresby, the local and international media, public opinion, government authorities and researchers often have a tendency to link gang violence with life in the settlements. The same can be said of Dili during the time of my field research. In the public discourse the violence was located in the 'hot spots' of the poorer neighbourhoods such as Bairo Pite, Becora, Fatuhada, Delta and Comoro. It was difficult if not impossible, for example, to convince taxi drivers to drive to some of the neighbourhoods in the daytime as they feared that their vehicles would be stoned; no taxis ventured out at night. As Higate and Henry (2009) note, international interventions often 'rezone' the areas where they are deployed based on security concerns about safe areas and 'hot spots', often in highly gendered ways. Dili was no exception, with most of the western parts of town being declared as unsafe (OCHA 2007).

In 1999, many Timorese (and internationals) I spoke with tended to dismiss members of the pro-Indonesian militias as either non-Timorese or from the lowest classes of society, of being stupid and violent, misguided thugs, and, though no one explicitly said it, male. A similar sentiment was evident later with respect to the gang/MAGs and RAGs violence in Dili. As the web blog Return to Rai Ketak (2006) recorded it,

> The Timorese diaspora, or educated class with good jobs here in Dili, tend to blame the *ema beik*, the stupid people who have come from the districts and live in these shitty peripheral neighbourhoods in Dili. Why do they resort to violence? They are ignorant. *Beik.*

Less pejoratively, numerous East Timorese NGO respondents in the background interviews also pointed to rapid urbanization as a key cause for the violence, as was also the case in the analyses carried out by Prüller (2008) and Streicher (2008). It is interesting that by and large it was educated people from middle-class backgrounds rather than residents of the poorer areas who raised this as a contributing factor. The latter, including members of MAGs and RAGs, on the other hand, tended to cast their neighbourhood in a good light, countering claims of the areas' purported instability and violence (see also Grenfell et al. 2009: 78–79).

The gang violence thus led to a stigmatization of the poorer areas of Dili as 'hotbeds' of criminal activity and their inhabitants as inherently prone to crime and violence. These neighbourhoods were therefore occasionally targeted by often heavy-handed crackdowns by the security forces. However, most residents of the 'problem areas' were not involved in crime and sought, where and when possible, to find regular employment, and they were more often the victims rather than the perpetrators of criminal activity. A lack of opportunities in the formal economy was, however, a constant source of frustration for many, and it came up recurrently in interviews with gang, MAG and RAG members (Myrttinen 2010; see also Wigglesworth 2013).

Moxham (2008: 2) sees the *bairos* as a site of both integration and disintegration where

> rather than Dili driving state-making through a process of... 'internal integration', it housed [in] it the failures to do so, where economic disintegration and political and violent challenges to the state were profoundly urban.

Kammen (2009, 391) has similarly argued that for the East Timorese political elite, the crisis 'revealed Dili's failure to be the site of modernity and rationality' which they had expected would arrive after independence. Scambary (2012) argues that the *bairos* have been the target of multiple waves of often partially circular migration from rural areas and have in part retained structures and networks based on those imported from, and connected to, the rural areas of origin, but have also seen the emergence of new networks and forms of coping, including links to MAGs and RAGs. Many of these networks, both old and new, are very much male-dominated, with older men drawing upon the physical capital of younger men to do their bidding, and women and girls playing a supporting role.

'The Hotel'

The space of 'The Hotel' stood in stark contrast to the squalor of 'The Camp' and the edginess of 'The Street'. The air-conditioned hotel – more specifically, the grand Hotel Timor[13] in the historical centre of Dili – was in many ways symbolic for me in terms of the disconnect between the 'expat bubble' and the reality inhabited by the vast majority of the East Timorese population. This bubble consisted largely of people whose job it was to help ameliorate the lives of the people literally less than 20 m away from the windows of the lobby, in the IDP camp, but who could have been on another planet or mere abstract figures. The IDPs, gang members and other East Timorese (often referred to collectively as 'these people') were discussed on the air-conditioned side of the tinted glass in a highly distanced manner, be it in casual conversation over *galãos* and fresh *pastéis de nata* or in PowerPoint-heavy presentations, and they were otherwise absent from this space.

While the Hotel Timor lobby, with its Suharto-era design, starched tablecloths and surly staff, which always took me back to the socialist-era Eastern European hotels of my childhood, with its bustle of internationals and insistence on a kind of lusotropicalism (even the bottled water was imported from Portugal in heavy, 0.2 l glass bottles, no expense spared), struck me as a particularly poignant manifestation of 'the bubble', there were other areas, too, where local nationals were only to be met as waiting staff, if even that.

Although the hotel catered to the international crowd as a whole, it tended to mostly appeal to a more upper-middle-class segment of the expats. Among the foreigners there were numerous gradations of class (reflected also in their employers – whether it was a UN agency or an NGO, or if you were a contractor), age and between nationalities. These gradations were reflected in which spaces one chose to relax and socialize in: some of the pubs catered mostly to middle-aged, t-shirted, working-class Australian contractors, 'stubbies'[14] of Victoria Bitter in hand (mostly but not exclusively male); others more to the more wine-oriented, upper-middle-class lusophone crowd (often older and slightly less male-dominated); and yet others to a younger, more international (but overwhelmingly white)[15] crowd of male and female junior staffers. Towards the end of Friday and Saturday nights, elements from these various scenes might converge, with some East Timorese as well, in the beachside discotheques in Metiaut and close to Pertamina pier, often a scene of violent, alcohol-infused hypermasculine stand-offs between Portuguese UNPOL and East Timorese police.

Each scene had its own hierarchies, be it UN pay grades (P1–P5), one's mobile phone number (which initially was a giveaway for how 'fresh' one was off the boat and/or seniority), knowledge of Tetum among some or, in some cases, how 'hard' a man (and occasionally woman) one was. In all of these hierarchies, to differing degrees, men tended to be in higher positions and also more visible in the various venues, be they the more prestigious restaurants or the late-night dives.

The outset of the crisis initially reduced mobility after nightfall to a bare minimum in Dili. This increased the importance of other forms of space for social get-togethers, such as dinner parties or more alcohol-oriented parties at expat homes (often with the hope of finding a sexual partner). As mobility slowly increased, there were numerous cases of sexual harassment of expat women by young East Timorese men (Bevan 2014). In all probability there were far more cases affecting local women given the high prevalence of sexual harassment in everyday life on the streets of Dili, but these remained for the most part unreported.

'The Karaoke Bar/Brothel'

I entered a fourth space of 'The Karaoke Bar/Brothel' when I volunteered to assist a local women's organization in the precrisis years in some research they were doing on sexwork in Dili, which dovetailed with the research I was doing on the unintended consequences of UN missions (Koyama and Myrttinen 2007). As the local female researchers were not able to enter these spaces without raising unwanted interest, I was asked if I could assist with some of the interviews with sexworkers and clients.

It was in this space that I started thinking more about the third category that was present in Dili in addition to the East Timorese and the expat *malae* – namely, the mainland and Southeast Asian Chinese and other Southeast Asians working in the city as employees, merchants and running businesses. Some of these people were building on connections going back to the Portuguese colonial era or the Indonesian occupation; others were far more recent arrivals. It is perhaps important to point out that while my focus here is on the 'dual use' space of the karaoke bar, the overwhelming majority of this section of the population was not engaged with this industry, be it as customers, facilitators or sexworkers. In fact, the people in these latter categories are very much the least visible section of an already less visible population.[16]

Although I had been interacting with this 'third category' of the population on a daily basis when frequenting supermarkets, Chinese- and Indonesian-run restaurants, renting equipment and so on, I had not put any real thought to their positioning *vis-à-vis* these other two broad groups of East Timorese and 'expats', a lapse which many of my fellow *malae* also committed. Although the visibility and vulnerability of this group rose as a consequence of the crisis, with many residents of Chinese and Indonesian descent evacuating themselves for fear of anti-Chinese/-Indonesian pogroms, and the Philippines government flying out its citizens lest they be 'mistaken' for Chinese or Indonesians, few international agencies, media representatives, researchers or NGOs picked up on the status of this section of the population.[17]

As part of the research into sexwork in Dili, which I repeated later during the crisis years at the request of a friend working for an international agency, I visited several karaoke bars suspected of being 'dual purpose', a suspicion that mostly did not take long to prove. The clientele were all male, and, apart from the most well-known ones, such as the Mayflower (which were also frequented, contrary to the 'zero tolerance policy', by UNPOL who parked their squad mini-van in front of the door), they were mostly made up of mainland and Indonesian Chinese, middle-aged men. For them, at least as far as I could tell, the available sex seemed to take a lower precedence than the sociability and bridging over of solitude and boredom by drinking lukewarm Chinese and Indonesian beer, smoking, belting out Chinese karaoke songs and socializing with other men in a similar position. Although I am quite certain sexual services were bought as well, it seemed that the social space of the bar/brothel as a place to hang out and relax, and to enjoy the camaraderie of other men who were also in the 'third' category, was more important to the customers.[18]

The women employed as bar maids/sexworkers whom I interviewed in these locations were all non-Timorese. During the first round of research in 2004, they were mostly sexworkers (in a broad sense of the word – i.e. not solely working as prostitutes) from Thailand, Indonesia and the Philippines (and a few from China), and sexwork was more explicitly marketed. During the crisis years, the market in Dili was almost exclusively in the hands of Chinese entrepreneurs who flew in women from mainland China, mostly Fujian, as well as some establishments with Indonesian workers. A parallel sexwork economy existed for male East Timorese clients with female East Timorese sexworkers, although wealthier East Timorese men, including high-ranking officials, also frequented the Chinese-run establishments, along with UN staff, openly flouting the official zero-tolerance policy on sexual exploitation and abuse (see also Koyama and Myrttinen 2007).[19]

Conclusions

What I have attempted to do in this chapter is to highlight how, within the relatively confined space of a small city, different parallel worlds coexist. These are partially co-related but in many ways are separate from each other, defined by gender, class and – for lack of a better word – 'race'. In many ways, but to differing degrees, all four spaces covered here are male-dominated, with women being in a less visible position but nonetheless essential in providing the reproductive labour that allows men to occupy and dominate these spaces. Class and 'race' define access to the spaces and to the social capital that are a necessary precondition of entry. These issues and dynamics are, of course, not exclusive to conflict- and disaster-affected societies and the humanitarian interventions which occur in them, but they do take on a particular pertinence as they are, on the one hand, mostly unacknowledged, while on the other hand they play a major role in defining how the situation is understood by the interveners (Smirl 2015).

In her work, Autesserre (2014: 174–183) examines the divide that exists (and is enforced) between local populations and interveners in what she terms 'Peaceland', the space in which peacekeeping and peacebuilding operations occur. She differentiates between voluntary and involuntarily separation, resulting from the human need to construct the 'other'; the difficulties of building links even when one tries; the cultural, social and economic barriers between the two groups; the transience of the interveners; and, in some cases, security concerns. All of these were present in Dili as well, and I would add some more factors: many expats were simply not interested in meeting locals, and vice versa. The differentiated spaces of 'The 'Hotel' and, for the non-*malae* foreign men (of course not for the sexworkers), 'The Karaoke Bar/Brothel' were also partially a 'backstage area', to use Goffman's (1959) term (*Rückzugsraum* in German) where one could withdraw from East Timorese reality and be in a comfort zone with other members of one's peer group, allowing for a partial separation of one's personal and work life (cf. Verma 2011: 75). 'The Street' and 'The IDP Camp' on the one hand and 'The Hotel' (and SUVs) on the other also formed what Verma (2011: 69) and Autesserre (2014) refer to as 'double fishbowls', where locals/interveners each gaze out of an enclosed space at the other.

Apart from providing useful metaphors for analysing the differences in terms of the lived realities and power imbalances between various groups inhabiting the same geographical area, applying a gendered, spatial analysis allows one to make visible spaces of particular insecurity and gendered, underreported security issues, such as sexual harassment, sexual and gender-based violence and sexual exploitation (e.g. in the case of mainland Chinese sexworkers or Southeast Asian migrant workers). A better understanding of the respective 'fishbowls' in which the various populations lived and the ways they viewed each other can also lead to a better understanding of the breakdowns in communication between local populations and those ostensibly there only to help them.

These breakdowns, or, more precisely, the lack of real communication in the first place, fuelled resentment and violence which in Dili discharged themselves in daily stonings of UN and other expat vehicles and occasional riots. At the time, the reaction of the foreign presence to the violence was, however, not to try to better understand the situation but rather one of a more forceful security presence and a broadening of the distance between the different populations of the city. An UNPOL vehicle that had just been stoned that drove past me one day provided me with what seemed like a particularly fitting visual metaphor: with its windshield damaged, the driver had placed his flak vest over the windshield, thereby blocking any view of the road ahead, and was stepping on the gas as hard as possible, oblivious to pedestrians on the street and to the fact that he was now in a safe zone, overreacting and not understanding the context. Without an understanding of what is going on around us and how we as interveners are willingly blocking our view through overly securitized approaches, we risk exacerbating the tensions with the people who we are supposed, at least in theory, to be assisting.

Notes

1. With some reluctance I chose the term 'racialized' as the politics of inclusion or exclusion in the setting I am describing were not based only on being East Timorese or not, because between internationals there were (and are) clear divisions between UN, INGO and other privileged foreigners and Indonesians, Chinese, and non-UN-/INGO-related Southeast Asians who would only be present in the *malae* (Tetum for foreigner, but with the connotation of 'white' foreigner) spaces as service personnel.
2. The original phrase, *anni di piombo*, refers to the period from the late 1960s until the mid-1980s in Italy and evokes, for me at least, the 'lead' of the bullets as much as the paralysing atmosphere of fear and mistrust in Dili during the *krize*.
3. These included both the UN Police (UNPOL), most visibly the Portuguese Guardia Nacional Republicana, the Australian-New Zealand military ISF and the Australian Federal Police.
4. For a humorous, self-reflective (if often painfully self-ironic) glimpse into the lives of 'internationals', especially of the junior grades, see http://stuffexpataidworkerslike.com/.
5. Tellingly, the Western and other privileged expats from whose perspective the classification is carried out are usually in the unnamed first person, while the local 'other' and 'host-country nationals' and 'lesser', non-OECD and non-white non-locals without the requisite financial or social capital are in the third category. See, for example, Chisholm (2014) and Higate (2012) on gender, nationality and 'race' among private security contractors.

6. This 'privileged' access was most clear when I was conducting research for a local women's rights organization on the sex industry because their local female researchers were not able to gain access to 'dual-use' massage parlours and karaoke bars (see below). Gender possibly, along with a host of other factors, also played a role when a female colleague and I interviewed male gang members separately, who to me would deny all acts of violence while (cautiously) glorifying them towards her.

7. Important exceptions are Autesserre (2014) and Smirl (2015).

8. Some people I interviewed, both nationals and internationals, suspected that this number might be inflated by people seeking to benefit from the aid given to the IDPs.

9. The Jardim camp was located in the very centre of the city, across the street from the Hotel Timor and from the port, and perhaps 500 m from the Palacio do Gobierno. Following the clearing of the camps, it was refurbished into a somewhat sterile park, with much of the bill being met by the notorious Jakarta-based gangster Hercules do Rosario Marçal, himself of East Timorese origin, who flourished thanks to ties to the Indonesian military but later allied himself, post-independence, with some of the leaders of the newly independent nation (Hyland 2008; Tatoli 2008).

10. The airport IDP camp's food stalls were *de facto* the 'eateries' and 'duty free shops' of the airport for a few months as the official ones had closed due to security concerns.

11. In spite of the massive international presence which should theoretically have been paying close attention to these issues, I have not come across any in-depth studies on sexual abuse, sexual harassment, or domestic, sexual and gender-based violence in the IDP camps, although there are bound to have been cases given the high prevalence in East Timorese society and the duration and scale of the displacement.

12. Information from my interviews with UNPOL in 2007.

13. It was only after having written the first draft of this chapter that I came across the excellent work of the late Lisa Smirl (2009), who uses the same hotel as one of her examples of an 'ideal' trope of the 'Grand Hotel' in an aidscape.

14. Australian beer bottle size of 330–375 ml.

15. In part, this can be attributed to cultural backgrounds (e.g. religious prohibitions on drinking, linguistic barriers, feeling as an outsider in the subculture of North American, European, Australian and New Zealand NGO staffers), in part to lesser purchasing power and in part to different priorities, such as using salaries for remittances to family back home.

16. At the time, the major mainland Chinese infrastructure projects had not yet commenced and the spatially different presence of the practically all-male, barracked workforce imported from China had not manifested itself. This category is also barely visible in the cityscape.

17. The exception to this would be organizations dealing with the trafficking of persons into or out of East Timor. Less problematic forms of migration to, and living in, Dili are underresearched.

18. My research on these bars was far more limited than in the other spaces, but during my visits there were no female customers.

19. I was not able to research this separate market, which caters to a mostly East Timorese clientele and employs local women. My one attempt to enter such a bar faltered when it was made quite clear that I, as a foreigner, was not welcome.

References

Autesserre S (2014) *Peaceland – Conflict Resolution and the Everyday Politics of International Intervention.* Cambridge: Cambridge University Press.
Ballard C (2002) The Signature of Terror: Violence, Memory and Landscape at Freeport. In David B and Wilson M (eds) *Inscribed Landscapes: Marking and Making Place.* Honolulu: University of Hawaii Press, pp. 13–26.
Bevan M (2014) Sexualised Tattoos and Street Assault – Navigating the Gender Dynamics of Working as a White Woman in the Global South. *International Feminist Journal of Politics* 16(1): 156–159.
Bovensiepen J (2009) Spiritual Landscapes of Life and Death in the Central Highlands of East Timor. *Anthropological Forum* 19(3): 323–338.
Chisholm A (2014) The Silenced and Indispensible – Gurkhas in Private Military Security Companies. *International Feminist Journal of Politics* 16(1): 26–47.
Direcção Nacional de Estatística (2009) *Timor-Leste Em Números, 2008.* Dili: Direcção-Geral de Análise e Pesquisa, Direcção Nacional de Estatística. Available at: http://dne.mof.gov.tl/upload/Timor-Leste%20in%20Figure%202008/Timor_Leste_in_Figures_2008.pdf (accessed 12 June 2015).
Durand F (2006) *East Timor: A Country at the Crossroads of Asia and the Pacific – A Geo-Historical Atlas.* Bangkok: IRASEC/Silkwork Books.
Goddard M (2005) *The Unseen City – Anthropological Perspectives on Port Moresby, Papua New Guinea.* Canberra: Pandanus Books.
Goffman E (1959) *The Presentation of Self in Everyday Life.* New York: Anchor Books.
Grenfell D, Walsh M, Trembath A, Moniz Noronha C and Holthouse K (2009) Understanding Community: Security and Sustainability in Four Aldeia in Timor-Leste: Luha Oli, Nanu, Sarelari and Golgota. The Globalism Research Centre. Available at: http://mams.rmit.edu.au/dfrtbpx26dgc.pdf (accessed 12 June 2015).
Higate P (2012) Martial Races and Enforcement Masculinities of the Global South: Weaponising Fijian, Chilean, and Salvadoran Postcoloniality in the Mercenary Sector. *Globalizations* 9(1): 35–52.
Higate P and Henry M (2009) *Insecure Spaces: Peacekeeping, Power and Performance in Haiti, Kosovo and Liberia.* London: Zed Books.
Hyland T (2008) The Gangster and Gusmao. *The Sunday Age,* 16 March.
International Crisis Group (2009) *Timor-Leste: No Time for Complacency.* Asia Briefing no. 87, 9 February. Dili/Brussels: ICG. Available at: http://www.crisisgroup.org/~ /media/Files/asia/south-east-asia/timor-leste/b87_timor_leste__no_time_for_complacency.pdf (accessed 12 June 2015).
Kammen D (2009) Fragments of Utopia: Popular Yearnings in East Timor. *Journal of Southeast Asian Studies* 40(2): 385–408.
Kingsbury D (2009) *East Timor: The Price of Liberty.* New York: Palgrave Macmillan.

Koyama S and Myrttinen H (2007) Unintended Effects of Peace Operations on Timor-Leste From a Gender Perspective. In Aoi C, de Coning C and Thakur R (eds) *Unintended Consequences of Peacekeeping Operations*, Tokyo: United Nations University Press, pp. 23–43.

Massey D (1994) *Space, Place, and Gender*. Minneapolis: University of Minnesota Press.

Moxham B (2008) *State-Making and the Post-Conflict City: Integration in Dili, Disintegration in Timor-Leste*. Working Paper 32(2), Crisis States Research Centre. Available at: http://www.isn.ethz.ch/Digital-Library/Publications/Detail/?ots591=0c54e3b3-1e9c-be1e-2c24-a6a8c7060233&lng=en&id=57417 (accessed 12 June 2015).

Myrttinen H (2008) Timor Leste – A Kaleidoscope of Conflicts. Watch Indonesia! – Information & Analysis, 1 April. Available at: http://www.watchindonesia.org/Kaleidoskop.htm (accessed 12 June 2015).

Myrttinen H (2010) *Histories of Violence, States of Denial – Militias, Martial Arts and Masculinities in Timor-Leste*. PhD Thesis, University of KwaZulu-Natal, South Africa.

OCHA (Office for the Co-ordination of Humanitarian Aid) (2007) Map of Dili Security Hot Spots as of 30 January 2007. Available at: http://www.mtrc.gov.tl/info/article.php?id=553 (accessed 12 June 2015).

Prüller V (2008) *The 2006 Crisis in East Timor – An Ethnic Conflict?* MA thesis, University of Passau, Germany.

Return to Rai Ketak (2006) Communication breakdown. In: Return to Rai Ketak. Available at: http://raiketak.wordpress.com/2006/09/ (accessed 12 June 2015).

Scambary J (2009) Urban conflict in East Timor. East Asia Forum, 18 September. Available at: http://www.eastasiaforum.org/2009/09/18/urban-conflict-in-east-timor/ (accessed 12 June 2015).

Scambary J (2012) Conflict and Resilience in an Urban Squatter Settlement in Dili, East Timor. *Urban Studies* 50(10): 1935–50.

Smirl L (2009) Spaces of Aid: The Spatial Turn and Humanitarian Intervention. In: *BISA Conference*, 15 December, Leicester. Available at: https://spacesofaid.wordpress.com/2014/03/12/spaces-of-aid-the-spatial-turn-and-humanitarian-intervention-draft/ (accessed 12 June 2015).

Smirl L (2015) *Spaces of Aid – How Cars, Compounds and Hotels Shape Humanitarianism*. London: Zed Books.

Stead V (2012) Embedded in the Land: Customary Social Relations and Practices of Resilience in an East Timorese Community. *The Australian Journal of Anthropology* 23(2): 229–247.

Stead V (2014) Homeland, Territory, Property: Contesting Land, State, and Nation in Urban Timor-Leste. *Political Geography* 45: 79–89.

Streicher R (2008) *The Construction of Masculinities and Violence: 'Youth Gangs' in Dili, East Timor*. MA Thesis, Freie Universität Berlin, Germany.

Tatoli (2008) Xanana/AMP fasilita Hercules loke bisnis iha Timor-Leste. Bulak ka beik? 29 September. Available at: http://odanmatan.blogspot.de/2008/09/xananaamp-hakarak-hercules-loke-bisnis.html (accessed 12 June 2015).

Teague M (1964) A Forgotten Outpost. *National Geographic Magazine* 37(1964): 110–123.

UN (2006) Report of the United Nations Independent Special Commission of Inquiry for Timor-Leste, 6 October. Geneva: United Nations. Available at:

http://www.ohchr.org/Documents/Countries/COITimorLeste.pdf (accessed 12 June 2015).

Verma R (2011) Intercultural Encounters, Colonial Continuities and Contemporary Disconnects in Rural Aid: An Ethnography of Development Practitioners in Madagascar. In: Fechter A and Hindman H (eds) *Inside the Everyday Lives of Development Workers. The Challenges and Future of Aidland*. Sterling: Kumarian Press, pp. 59–82.

Wigglesworth A (2013) The Growth of Civil Society in Timor-Leste: Three Moments of Activism. *Journal of Contemporary Asia* 43(1): 51–74.

6
Local Agency in 'Global' Spaces? The Engagement of Iraqi Women's NGOs with CEDAW

Annika Henrizi

Introduction

The fight for women's rights in different regions of the world has given rise to a process of 'global norm creation' (UN Women 2009). CEDAW is a milestone in this process, which aims to improve the reality of so-called 'local' women in various countries. This very distinction between the 'global' and the 'local' has become commonplace in the practice and discourses around women's rights. From a critical feminist perspective, global instruments are regarded as hegemonic and distant from local realities. The same argument is prominent in critical approaches to peace and conflict studies where critics of the so-called 'liberal peace' argue that the promoted forms of peace follow definitions and preferences of the global North, and that local perspectives and practices are neglected and agency withdrawn from local actors.

By dealing with Iraqi women's engagement within the CEDAW process, this chapter is located at the interface between both discourses. I take the distinction between the global and the local as a starting-point to analyse if and how Iraqi women have acted as capable agents during the CEDAW process. My observations on the engagement of Iraqi women's NGOs (the local) within the CEDAW process (the global) reveal that a binary reading of the global and the local does not match realities on the ground. Rather, aspects of global and local are entangled throughout the process and are hard to separate.

Feminist scholars have challenged the global–local divide by showing how CEDAW is linked to various notions of space (national, transnational, global and local) and by referring to different sets of actor and of space (see Caglar et al. 2013; Zwingel 2012). For instance, Zwingel (2005) elaborates how a transnational perspective can be applied in the CEDAW context, focusing on what she calls interrelatedness. While many of these approaches concentrate on the process of norm distribution or translation, of interest here is how CEDAW is negotiated between actors in different spaces. I see it as a potential space for interactions between (so-called) global and local actors. In the context of this chapter, I focus on the involvement of NGOs and analyse whether local agency is possible in global spaces.

Both the global and the local connote a spatial perspective, but neither orthodox nor critical approaches to conflict studies discuss their understanding of these spatial markers. I argue that the failure to recognize the potential for local women's agency in global spaces is a result of a binary reading of space. By rethinking the underlying concepts of space, we are able to gain a deeper understanding of how space and agency are intertwined. Approaches that regard the local and the global as being opposed to each other build on an absolute understanding of space that remains confined to dichotomist thinking. Space, in this sense, is a container in which action takes place. By contrast, a relational reading of space allows us to see the duality of acting and space. Spaces are – following Löw's (2001, 2008) take on relational space – an arena in which actions take place, but at the same time they are constituted through actions and are in principle prone to change. Hence, if spaces – including the global and the local – are constituted through actors, they cannot be absolute but rather have the potential to interrelate and overlap.

After discussing how the binaries of the global and the local can be deconstructed by a relational concept of space, I move beyond the 'global versus local' debate by focusing on the spaces in which CEDAW is negotiated. Negotiations take place in spaces, and these spaces are more than the context in which action occurs. Rather, the material and social constitution of those spaces shape the very outcome of the process, for power relations between actors are reflected in spatial settings. At the same time, actors take part in the constitution of space so that the opportunity to position oneself within a space can be a powerful tool for agency.

To analyse the involvement of NGOs in CEDAW from a spatial perspective, I combined participant observation, semistructured interviews and textual analysis. As a participant in both the shadow report training for NGOs and the official dialogue between NGOs, the Iraqi Government and the CEDAW committee in Geneva, I could observe how spaces were divided and organized (Barrasi 2013: 56), and how hierarchical structures and power relations were reflected in spatial settings. Also, during these meetings I gained insights into how actors found opportunities to position themselves in spaces, thereby potentially exercising agency.

This chapter proceeds by sketching a relational approach to space as proposed by Löw (2001, 2008) that allows the linking of agency and space on a theoretical level. I then briefly introduce CEDAW and the actors involved in the Iraqi process. My empirical analysis starts with insights into interactions between Iraqi NGOs and International Women's Rights Action Watch (IWRAW) prior to the CEDAW session in Geneva. These processes laid the groundwork for potential agency and are important in understanding the negotiations that took place in Geneva. I then examine how NGOs acted and interacted during the Iraqi session in Geneva from a spatial perspective. Lastly, I conclude by summarizing the empirical outcome and discuss theoretical implications for researching local agency.

A relational view of space

The notion of space and its conceptualization has long been neglected in the social sciences. Space has mostly been understood exclusively as a material, absolute entity. By contrast, both sociologists and critical geographers advocate for an understanding of space as relational (see e.g. Berking 2006; Löw 2001, 2008; Massey 1994, 2004). The value of a relational understanding of space for this chapter is two-fold. First, it helps one to move beyond the dichotomy of the global and the local, and, second, it allows one to link actors and space from a theoretical perspective, thus providing a way to analyse agency from a spatial perspective.

Drawing on the duality of structure and action proposed by Anthony Giddens, Löw developed a relational conceptualization of space that integrates action as a crucial element. Space in this sense is understood as a 'duality of structural ordering and action elements' (Löw 2008: 25). Spaces come into being through the combination of 'spacing' and 'synthesis' while at the same time recursively affecting processes of action. Spacing means the creation, formation and positioning of social goods and human beings; these processes always take place in relation to other processes of positioning. Through synthesis, humans and social goods are perceived, ideated and recalled, and thus integrated to form a space. Spacing and synthesis happen simultaneously and continuously; space is constituted in the course of everyday life. For our purposes, it is important to keep in mind the duality of action and space that is emphasized in Löw's conceptualization of space: 'Spaces are not only the point of reference for action or the product of action but, as institutions, also structure action' (Löw 2008: 33). Space in this sense is a platform on which agency unfolds while it is constituted through action at the same time.

Having outlined the concept of space that I will use for my analysis, I shall briefly reflect on the differentiation between place and space, which is heavily debated in both sociology and geography. Since I use Löw's concept of space for this analysis, I mostly refer to space rather than place. Yet place remains important for the constitution of space: as we have learned, space incorporates both social goods and human beings. Place, on the other hand, denotes an area that can be 'specifically named, usually geographically marked' (Löw 2008: 42). Places are a crucial part of the constitution of space; in daily life they are hardly ever separated from the people and objects that are attached to them. Together, these elements are perceived – and thus integrated – as space (Löw 2008: 42). Contrary to a binary conceptualization of place versus space, this reading allows for considering space and place to be inextricably bound together. Place and space are difficult to distinguish not only in daily life but also in academic analysis. To address that difficulty and to be clear to the reader, here I refer to space when I mean its social constitution (which includes the geographic place) and to place when referring primarily to a geographical area.

The mutual entanglement of space and place is important for our reading of the global and the local. The local is not to be equated with place, and neither the global with space. Instead, the local can refer to a geographical place as well as to a space that incorporates people, places and goods. As with all other spaces, the local is constituted at a certain point in time; it can never be static or homogenous since people are part of its constitution. The global is often described as a fluid space without any attachment to place. Yet it is constituted by and negotiated within certain places and local spaces that shape its very nature (Berking 2006: 13–15). Even though the constitution of the global is not the focus of this chapter,[1] I will refer to it briefly in the context of CEDAW when appropriate.

Let me briefly return to the question of power. The deep entanglement of space, gender and power has been central to feminist studies for a long time. In this context, Ruhne (2011) develops a conceptualization that draws on both space and power as relational and in which she defines power as an interdependent relationship that is always relational. Power is not 'a thing' – something we can or cannot possess – but a result of negotiation and is hence relational (Ruhne 2011: 157–170). Spaces – like all other structures of society – are inevitably structured through power relations, yet at the same time they shape power relations, and they can be both enabling and disabling for certain actors or groups of actors. In a positive sense, power also refers to agency so that agency, understood as the opportunity to act, is the result of power relations. Hence agency is not static but permanently negotiated (Helfferich 2012: 9). When we link agency and power to our concept of space, it follows that the possibility of individuals to act, to position themselves and social goods (spacing), and thus to constitute or change spaces, is highly dependent on power relations.

For Ruhne (2011), gender is a structuring element in any space because all spaces are gendered and power is negotiated – although not exclusively – along gender divisions. Spaces can be open or closed, empowering or not, in different ways for men and women. In this sense, CEDAW is exclusively designed to impact on women's matters in how the spaces in which it becomes relevant are organized and how women as actors find opportunities to enter and position themselves within these different spaces.

Before I move on to a short introduction to CEDAW and my empirical analysis, let me briefly summarize how I use the relational concept of space and how this adds to a deeper understanding of the observed reality. If space is constituted through actions, and hence in principle is prone to change, it cannot be homogenous or static. If we regard the local and the global as constituted spaces, the global and the local are not to be separated because they depend on the actors involved. As I will elaborate below, spaces can even support notions of the global and the local at the same time. Furthermore, they are perceived through synthesis. I focus here on the process of 'spacing', but it is important to acknowledge that different actors might perceive spaces differently and that this informs their actions. In the following empirical analysis of spaces of interaction, I elucidate how spatial constitutions reflect power relations and influence opportunities to act. Likewise, I aim to show how local actors position themselves and social goods in global spaces (spacing) and thereby become capable agents.

CEDAW as a global instrument for women's rights

The UN General Assembly adopted CEDAW in 1979, which was regarded as a milestone in global norm creation in the field of gender equality

(UN Women 2009). Although the convention started as a classical inter-governmental regime, it has progressed to a 'transnational network enforcing women's rights'. It is now considered an important tool to hold states responsible for discrimination against women in all spheres of public and private life (Zwingel 2005: 400).

Different groups of actors play relevant roles in the outcome of the process. States that have ratified the convention are obliged to submit national reports every four years on the measures they have taken to ensure compliance with the treaty's provisions. They do so by submitting a written report and engaging in what is called a 'constructive dialogue' with the committee – consisting of 23 international experts on women's rights – during one of its sessions in Geneva or New York. As a result, the committee issues recommendations that the state should work on until the next report, and even identifies two pressing issues that have to be addressed within two years (Zwingel 2005: 404).

Local NGOs have the opportunity to submit an alternative report (called a shadow report) in which they offer their assessment of state compliance as well as data on the most pressing issues for women in the country. Also, they are invited to the sessions in Geneva or New York to present their case and engage in a dialogue with the committee. This is an opportunity to influence the outcome of the session and the committee's recommendations in particular (UN Women 2007). As my research reveals, the dialogue between NGOs and committee members is not restricted to the more formal activities (the written report, a formally planned lunch meeting and the speech within the session itself) but also takes place informally between sessions.

Another relevant actor is IWRAW Asia Pacific, which was founded in Kuala Lumpur in 1993. Although the organization helped to promote and raise awareness about international women's rights at its outset, one of its main tasks has since become to encourage local NGOs to participate in the CEDAW process and to support them in this. As such, it cooperates with the CEDAW committee in order to assist local NGOs both in practical matters (e.g. the submission, printing and delivery of reports) and in more substantial aspects of how to channel their activism within the CEDAW framework – for example, how to draft a report in compliance with CEDAW, how to prepare a statement and how to develop follow-up strategies (IWRAW 2010b; Zwingel 2005). IWRAW's Global to Local programme aims to fill 'the gap between human rights monitoring by the CEDAW Committee at the international level, and grassroots activism of NGOs demanding government accountability at the national level' (IWRAW 2010b).

In the Iraqi case, the state report debated in 2014 was the first since 2000 and therefore the first since the fall of the regime of Saddam Hussein and the US occupation of Iraq. This made Iraq a case that drew much attention not only from committee members but also from other international agents working in Iraq (NGO leader and IWRAW trainers, 2014, personal communication). In their report and during the session in Geneva, Iraqi Government officials pointed out that they were experiencing an extremely difficult situation, given that the country faced (once again) a new outbreak of violence and a deteriorating security situation (Iraqi state report 2013: 44).

The NGO coalition of CEDAW shadow report on Iraq involved three different network organizations with various women's NGOs as members (Shadow report 2014: 1). Most of the organizations operated out of Baghdad. However, for the purpose of CEDAW the task was to form a coalition that brought together Baghdadi NGOs with those of other regions – especially the Kurdish part of Iraq – and submit a joint shadow report. Officially their task was to support the independent experts with their specific knowledge of the local situation (Zwingel 2005: 412). IWRAW supported the NGO coalition throughout the process. A training session in Beirut was conducted in cooperation with a local NGO in order to prepare the shadow report, and the NGOs regularly contacted IWRAW for further advice. Ultimately, the NGO coalition took part in the programme, which was held in Geneva prior to the Iraqi session.

Negotiating local agency

Before I move to the core part of my analysis, which deals with the interactions in Geneva, I will briefly sketch some observations from the shadow report training in Beirut mentioned above. Some of the friction one might expect to be part of the negotiation process had already become apparent during the training. Some of that can be framed through a local versus global lens, but other issues arose between local actors or were related to different and conflicting paradigms. The insights from Beirut add to a deeper understanding of local actors and their perceptions of and positioning towards CEDAW. As stated earlier, agency is not something one can possess but something that is always negotiated. The process of negotiating local agency had already started in Beirut (or even earlier) and paved the way for what happened in Geneva.

An orthodox, binary reading of global and local assumes that the local is a sealed, uniform entity (Henrizi 2015; Mac Ginty 2011). However, the 'local' as such does not exist; notions of the local coexist and compete. The local here includes the geographic place of Iraq (and/or different Iraqi regions and cities) but also refers to space, including people and their participation in the formation of space. The NGOs present in Beirut represented only a small part of Iraqi society and Iraqi women; they were only one specific manifestation of a local reality. What is more, even this group of NGOs is very diverse in their place of operation as well as in their aims and concerns, not to mention in terms of their access to power: the training was organized and mostly paid for by Iraqi Al Amal (one of the most well-known Iraqi NGOs), but members of other NGOs[2] were also invited to join (Al Amal member, 2013, personal communication). Yet power relations emerged from the financial and organizational efforts that Al Amal had made, as its employees assumed a leading role throughout the training. They tried to be inclusive by inviting and paying for members of other NGOs to come to Beirut but, according to a conversation with one member (Al Amal member, 2013, personal communication), they did not have the capacity to invite members of other NGOs to Geneva. Financial resources as one aspect of power therefore determined whether activists were given opportunities to position themselves in a 'global' space. Neglecting the diversity of the local thus reinforces the binary reading of global and local.

Within the CEDAW process, the diversity of the local laid the groundwork for friction between the diverse local realities and priorities, and the need to unify one's efforts for the sake of CEDAW: the differences in the women's situations – especially between the Kurdish part of Iraq and the rest of the country – had to be included in the joint report. Beyond the report itself, cohesiveness was, as IWRAW trainers pointed out, essential to be successful in front of the committee:

> Please, ensure that everyone is on the same page, refers to the same evidence and that you have agreed on the underlying reasons for the problems you raised. NGOs have to have a comprehensive coherent approach!
>
> (IWRAW trainer, 2013, group conversation)

Hence to exercise power among the committee and in one's own state (which is ultimately what it was all about), activists had to leave behind differences and disagreements. The agency of local NGOs was restricted here because they had to follow the logic of CEDAW.

Friction also arose between Iraqi NGOs and IWRAW trainers. Although IWRAW's engagement is meant to empower NGOs (Get

Involved 2013), it also restricts agency at times. On its website, IWRAW claims to operate internationally and to be experienced in lobbying human-rights treaties such as CEDAW in national contexts (IWRAW 2015), which is linked to certain forms of power. Being an NGO that operates internationally means having knowledge of many localities throughout the world (through contacts with local NGOs) but also having knowledge of CEDAW as a global instrument for women's rights. Throughout the training, IWRAW trainers used examples from other countries to show how NGOs had managed to fit their perceptions and needs into the CEDAW framework (Henrizi 2013: 3–5). They can be considered translators of the 'global', as they are able to speak the 'language' of the UN, and they have a great deal of experience in dealing with local NGOs and transferring knowledge. This advantageous role led to a dispute during the workshop. When, on the second day, Iraqi participants proposed adjusting the training programme to fully meet their needs, IWRAW trainers refused to do so, arguing that they knew better: 'You know, we have experience with that' (IWRAW trainer, 2013, group conversation). Iraqi reality mattered for them when it concerned the issues that should be introduced in the report but not when it concerned the general strategy and methods of the training. The conflict above can be viewed through a global versus local lens, in which the global NGO – as the more powerful one in the group – withdrew agency from local activists. Local agency was restricted here, and some activists were concerned about that, as they confided to me after the session. However, as I learned through informal conversations after the discussion, they rather interpreted it as a need to compromise, which they were used to doing when working with different partners, especially from other countries (NGO members, 2013, personal communication). What helped to overcome fractures was a shared identity as women activists, which was acknowledged and strengthened during the breaks and over dinner. IWRAW and NGO members shared their stories and challenges as women activists and showed mutual respect for their work (participant observation, 2013).

Another occasion where agency was restricted or predetermined was in the drafting of the report. During the training, the trainers emphasized several times that the structure and language of the shadow report had to follow the protocol of the CEDAW convention and refer to the forms of discrimination listed by CEDAW (Henrizi 2013: 8–12; extensive guidelines on how to organize and write a report can also be found on IWRAW's website (IWRAW 2010a)). In the end, the process was about classifying the local reality and documenting the detailed knowledge of the local space (violations of women's rights, instances of discrimination) according to the format prescribed by the convention. Many stories had to be cut short for the sake of time and possibly because they were not relevant to the outcome. Iraqi participants were concerned about this at the time but framed it in terms of the usual bureaucratic hassle with UN agencies, something they were used to in their everyday work: 'When they were talking about individual demands from the people, it's a very, very paralyzed process, for individuals to reach the committee' (NGO member, 2013, personal communication). NGO members did not regard it as a dispute between the global and the local but between huge bureaucratic institutions and the priorities of practitioners who deal with individual cases.

My observations from Beirut show that disagreements between actors exist and that a romanticized view of how the global (CEDAW) is empowering the local is misleading. Power hierarchies become visible in interactions, and indeed shape and restrict opportunities for agency. Yet CEDAW promised to be a helpful instrument for local NGOs. I shall now discuss how local agency was finally negotiated in Geneva.

Interacting on global and local grounds

Lunch meeting – Geneva, building E, room 3008

I have elucidated how grounds for local agency within the CEDAW framework were negotiated in Beirut and how notions of local and global emerged during the training. As I explained earlier, I aim to examine the Iraqi CEDAW process in Geneva from a spatial perspective, which will allow us to see how agency and space are intrinsically linked. In the following section, I will focus on two spaces that shaped the outcome of the process while at the same time providing an arena in which local agency could be negotiated: room 3008, building E, UN Office in Geneva; and the Convention Hall, UN Office in Geneva. These were the spaces in which the lunch meeting (supported by IWRAW's Global to Local programme) and the official session for Iraq (57th Session) (OHCHR 2014) took place.

Before I begin to analyse interactions and spaces, I shall briefly reflect on how global and local refer to those spaces. The Global to Local programme is meant to adjust local NGOs to the UN environment (with respect to content but also to the place, building and atmosphere itself), and help to prepare the lunch meeting and session itself. The title of the programme suggests that CEDAW is global, but it is also local in different ways. CEDAW evolved as a product of Western, intergovernmental negotiations (Zwingel 2012). Beyond this evolution, it is translated into and implemented within local spaces. And, as I will explain below, it is also local because CEDAW is negotiated and applied at the (indeed) local space of the UN Office in Geneva. In fact, the task of adjusting NGOs to the UN environment can be better described as adjusting them to a very specific local space (Geneva and the office proper). Yet, as I entered the space, I also perceived signs of globality that I will detail below.

Let me start with the spatial setting of room 3008, where the preparation for the lunch meeting and the meeting itself took place. Its spatial constitution includes place, material and social goods, and human beings. Room 3008 is part of building E at the UN Office in Geneva. The compound consists of various modern, Western buildings that one could find anywhere in the Western hemisphere. As such, it is a local space. However, the national flags of member countries that decorate the compound as well as the international staff moving about constitute the space as a global space. The room itself looks (to me) like an ordinary, rather boring meeting room in a huge conference building, equipped with chairs, tables and technical equipment (participant observation, 2014). The room was assigned to Iraqi participants for the duration of their work in Geneva. While the place of the room and its furniture were predetermined, other parts of the spatial constitution were subject to change. When I entered the room, it was already 'occupied' by Iraqi participants. People were placing sweets and tissues on the tables, and spreading their working materials all over the place, making it more similar to the Iraqi meeting rooms I visited when taking part in NGO meetings in Iraq (participant observation, 2014).[3] What might seem trivial at first sight is, if we reconsider how spaces come into being, an important part of the constitution of space e.g. the room and furniture) were fixed, NGO members managed to shape the constitution of the space into an Iraqi space by positioning social goods. That might have helped them to feel more present, more comfortable and more efficacious. For a while, the room in which a global framework was negotiated, and which was located in Geneva, Switzerland, became – through the process of spacing – a space for local Iraqi women to become potentially powerful actors.

The Iraqi delegation and committee members interacted during the lunch meeting. Some 17 out of the 23 members of the committee showed up at the NGOs' invitation, which was seen as an initial success, given that attendance was voluntary. The Iraqi participants were very excited by the attendance and kept mentioning the number throughout the day: 'We got 17. Imagine, 17!' Reflecting on the meeting afterwards,

IWRAW staff members also cited the number as a great success that indicated strong interest in the Iraqi case (IWRAW trainers, 2014, personal communication). From a theoretical point of view, I argue that the positioning of committee members in relation to other human beings – namely, NGO members – gave the latter the opportunity to make their voices heard by as many potential allies as possible. The spacing of many – who literally filled the room – changed the constitution of the room as such. Hence the room eventually became a powerful space for agency.

After a short statement by the NGOs on the most pressing issues for women in Iraq, committee members were keen to ask for information and clarification. Given that there continues to be little public knowledge about the situation of women in Iraq, committee members asked for advice about what to focus on and what to question during the session: 'You have to tell us about the situation, you have to tell us what to ask the state' (committee member, 2014, group conversation). The aim of the committee was (in the long run) to improve the local situation of women about which they had little knowledge. By contrast, being from the local space meant having knowledge of Iraq and proved to be grounds for empowerment in this case. The NGOs were to act as intermediates (Schlichte and Veit 2007) between two different spaces: Iraq and room 3008 in Geneva. The NGO members were very satisfied and enthusiastic in their discussion after the meeting. During supper they discussed how they imagined committee members might question their state representatives and which of the members they considered to be most promising for their case. One of their aims had been to draw as much attention as possible to the situation of Iraqi women:

> We have to make use of the chaos [the new outbreak of violence] and the attention it brings to Iraq. Internationals and diplomats in Iraq are very interested in the outcome of CEDAW.
>
> (NGO member, 2014, personal communication)

They felt they had succeeded because committee members participated and showed interest. NGO members had been assigned a (very restricted) space and time in which to position themselves and advocate for their concerns. From an analytical perspective, the positioning and action of committee members in that very space and at that time potentially enhanced the local agency of women in Iraq.

However, NGO members made clear where their agency towards the Iraqi state was restricted and how the committee could act to assist them: 'We need your help to pressure the government and get information' (NGO member, 2014, group conversation). The Iraqi government frequently forms committees, creates working groups or reports on women's issues without transmitting information about the procedures and consequences of these structures. When the lunch meeting took place, it was, for instance, unclear to the Iraqi participants what had happened to a draft of a national action plan (NAP)[4] in the context of UN Security Council Resolution 1325 and a plan to amend the Iraqi penal code (participant observation, 2014; see also Shadow Report 2014: 12). During the session the committee was asked to question the state about these very issues.

From a critical perspective, one might argue that power relations between the global and local persisted in CEDAW and that these were even reflected in the spatial setting and the place of the meeting. As such, the setting was far away from local Iraqi reality. Power hierarchies structured action, as the times of the meetings and part of the spatial setting were predetermined. Yet if we remember that space comes into being only through actions, and that spaces are in principle subject to change, local agency is also possible in spaces that are predominantly global. On the basis of my observations outlined above, the Iraqis themselves perceived the space as one that was enabling for them as women rather than a hegemonic global space. The global framework of CEDAW promised to restore agency towards the Iraqi state in a situation in which opportunities to act were limited by the state's behaviour.

Before I move on to analyse the official Iraqi session, I shall note another occasion that reveals how the global and local were intertwined through the committee members. The interactions between NGO and committee members continued after the meeting in many informal ways and in different places within the office complex, only some of which I was able to observe, whereas other interactions were recounted to me later. NGO members were keen to lobby committee members between sessions, which meant that they took every opportunity to advance their priorities for the session. While the committee itself was presented as global in nature, consisting of experts from 'around the world', as the UN itself puts it (OHCHR 2015), the lobbying process revealed that its members were also perceived as locals. The place of origin was one important foundation on which coalitions could be built. For example, the Iraqi NGO members focused in particular on a Lebanese committee member, assuming that being geographically (and culturally) close would have a positive effect on their ability to understand the problems of the Iraqi case. Yet alliances were not only forged along geographic lines but also based on other aspects of social identity – for example, the professional expertise or interest of committee members. It was important to the NGO members to find out which of the committee members shared an interest in the issues that were most pressing in Iraq and therefore would likely take a stand for the 'right' ones (family law, violence against women, women in conflict) (participant observation and personal communication, 2014).

The Iraqi session – UN Convention Hall, Geneva

The 57th session of the CEDAW committee, which included Iraq, took place in the Convention Hall of the UN Office in Geneva. I have already described parts of the spatial setting – namely, the geographic place and the complex at the UN Office. The hall itself was a rather fancy conference room with padded seats and a carpeted floor, which I felt created a serious, almost ceremonial, atmosphere. Within the Convention Hall, different spaces were assigned to the various participating groups. A closer look at the seating arrangement revealed that power relations as well as bases for interaction were reflected in the spatial organization. The front stage hosted the session chair (and her administrative team) as well as the head of the delegation of the Iraqi state – namely, the minister for women's affairs. In the centre of the hall, committee members were seated in a semicircle. NGOs and the representatives from the Iraqi state were seated in the rows behind the committee, each on either side.[5] The remaining space left room for international NGOs and other observers. The NGO delegation from Iraq was quite large (compared with other NGO coalitions) and therefore took up many seats. If we regard spacing – the positioning of NGO members in relation to others – as an important part of agency, local NGOs became powerful through their participation in the constitution of space. Still, the government delegation was considerably larger. It consisted of different experts on the various topics up for debate (participant observation, 2014). The spatial arrangement reflected power relations between the Iraqi Government and NGOs because the government had more financial resources to send its members and experts to Geneva. The overall spatial organization of the hall reflected the interactions that were supposed to take place. The Iraqi delegation – primarily the head of delegation – was the focal point of the session. It was they who were given time to present their report and who were also called upon to answer questions posed by the committee. From my observations, the seating arrangement of the NGOs gave the proceedings a sense of their backing the committee – and/or the committee speaking also on their behalf.

While the dialogue between the state and the committee took most of the time, the Iraqi NGOs were formally assigned ten minutes to read out their statement, but that was not the only way for them to ensure that they were heard. Perhaps even more essential was whether the committee members asked the 'right' questions, identified the 'weaknesses' in the state report and drew the 'right' conclusions. Having the committee on their side was therefore even more crucial for the NGOs in exercising power towards the state. The comments and whispered conversation of NGO members between sessions showed that they were satisfied with the questions the committee posed (and the evasive answers of the state). They felt a sense of power upon hearing their own concerns discussed in this very space and time (participant observation and personal communication, 2014). The spacing and action of committee members towards the Iraqi state promised to be empowering for local NGOs: their concerns were spoken aloud by actors who were perceived as rather powerful and within a space that reflected the gravity of their concerns. Whereas their own agency towards the state was limited in Iraq (state representatives simply refused to give answers and meet requests), their moving to and positioning themselves in another space helped to hold the state responsible – at least in word. Being in the same room obviously does not mean that two parties hold equal power; yet here it left the government unable to avoid confrontation and (given that the session was recorded) unable to ignore any parts of the ensuing discussion. In that sense, what took place during the session could also be seen as a competition between a government and NGOs over the truth about the Iraqi space and women's local reality. Alliances (as perceived by the NGOs) were not built along the global–local divide. Being women activists (both locally and globally active) again proved to be of greater importance than geography because NGO members assumed that committee members were on their side as far as the substance of the priorities was concerned (NGO members, 2014, personal communication). CEDAW and the Convention Hall as the space in which it manifested at a certain time became an empowering space for Iraqi women through their being in a particular place and their concerns occupying that space. It was a women's space as constituted by the positioning of people (who were mostly women) and as structured by the power that was ascribed to women. Power hierarchies between men and women that are dominant in Iraq did not prevail there. While in Iraq spaces are often closed to women, the UN complex was a space that was very open for women during the time of the Geneva sessions.

I have discussed how CEDAW as a so-called 'global structure' might prove to be enabling for local actors and how the interactions and power relations constitute, and are shaped by, the spaces in which they take place. The question of local agency in the context of CEDAW cannot be answered definitively here or elsewhere because the convention is an ongoing process. At any rate, I have shown moments in which agency unfolds. Before I conclude, I shall shed some light on current developments following the session in Geneva.

The Geneva session ended on 18 February 2014. After travelling back to Iraq, the NGO members waited for the committee's concluding observations that were supposed to be issued within a fortnight. Once back, the activists were immediately confronted with the very same issues that they had been discussing in Geneva: despite the outrage and protests of many Iraqi women, Jaafari law was passed through the Council of Ministers at the end of February 2014.[6] The possible amendment of the law and its restrictions on women's freedoms had been one of their main concerns in the shadow report and their oral statement (CEDAW 2014; Shadow Report 2014, annex 4). Members of the CEDAW coalition passed the information via e-mail to committee members to keep them updated: 'as for the Jaafari law, since it was cleared yesterday by council of ministers, we managed to contact the CEDAW committee, send them details, so that they can emphasize on the issue in their concluding observations' (NGO member, 2014, personal communication). Also, they used CEDAW as a basis to further strengthen and legitimize their protests against the law:

> in spite of confirmation from State delegation to International CEDAW committee members last February, that the sectarian Jaafari personal status law and Article 41 in the constitution are suspended to be deleted soon, the Council of Ministers passed the law.
>
> (NGO member, 2014, Facebook post)

Finally, the members of the NGO coalition managed to have many of their main concerns included among the committee's concluding observations (COs), among them the Jaafari law and the NAP per resolution 1325: 'It [the committee] calls upon the State Party to establish a clear time frame for finalizing the draft national action plan to implement Security Council resolution 1,325 ... The committee recommends that the State Party ... Immediately withdraw the draft Jafaari personal status law' (UN 2014: 3–5). Those two issues even became pressing ones for the committee, meaning that the government has to work on them according to a two-year framework (UN 2014: 17). Powerful influence is not limited to situations in which it can be traced back using 'hard facts', and this marks one occasion when actors claimed to be powerful and perceived themselves as such. The NGOs positioning themselves in the 'global space', or rather at a certain place in Geneva, was seen as a tool for them to gain power in the local space, to influence political and social practice in Baghdad and Iraq at large:

> We find 1325 and CEDAW very essential to hold Iraqi government obliged to women's political participation, negotiation and peace talks. It's a must for all of us activists in and outside Iraq. COs gave us a reference to hold government obliged.
>
> (NGO member, 2014, personal communication)

On a broader level, power relations were not in question as opportunities to act were restricted to the scope of action predefined by CEDAW. Yet acting within a global space remained a promising option for NGO members. During an interview and informal conversations in the autumn of 2014, two NGO members pointed out that they continued to work on creating media awareness about CEDAW and its outcome for Iraq. They were also about to become engaged with another instrument of the UN, the UN Security Council Resolution 1325. Hence they continued to plan to actively engage in global spaces.

Conclusion

The insights and empirical findings about CEDAW do not allow me to draw any final conclusion because the process is by its very nature a work in progress. For the NGOs it remains to be seen how powerful the use of CEDAW proves to be in effecting the actions of Iraqi state representatives. Instead of drawing a wider conclusion, I will briefly reflect on the empirical outcomes of my research and discuss the theoretical approach as well as its benefits in advancing a critical agenda in peace and conflict studies.

My observations have shown that notions of 'global' and 'local' do exist and shape the interactions of NGOs in the context of CEDAW, yet 'global' and 'local' are not to be understood in a binary way. I have elaborated how CEDAW as a process of negotiation between different actors is much more complex than a binary reading of local and global suggests. Instead, if we consider spaces to be constituted by people, neither the global nor the local is static but is in principle subject to change and – depending on the actors – potentially overlaps with the other. The social space in which negotiations take place is structured by power and power imbalances, and not only along the lines of global versus local.

What further added to my understanding of the involvement of the NGOs and the power relations throughout the process was to regard agency and space as being closely linked. As local actors, Iraqi women's NGOs managed to literally enter the space of the UN in Geneva and position themselves and their concerns in different ways. From their perspective, their involvement was a great success, especially with regard to the interest of committee members in their concerns and the concluding observations. As already mentioned, how these achievements will be put into practice by the Iraqi state will be borne out in the coming years. The question remains as to whether or not the agency that the NGOs exercised in Geneva will also be successfully negotiated in the space of Iraq. Only then would local agency be enabled through spacing in so-called 'global spaces'.

A relational conceptualization of space has proved fruitful in analysing encounters between so-called 'local and global actors' in the context of CEDAW. My empirical analysis shows that a binary reading of the global and the local, of space and place, fails to match reality on the ground. Regarding CEDAW as a process of negotiation that is constituted in various spaces and places at different times potentially broadens our perspective on power relations that are at stake in these encounters. That is to say, power is reflected in the organization and division of space that influence opportunities to act, and at the same time power is exercised by different actors who constitute or reconstitute spaces through their process of spacing. To include spaces and places when taking a critical perspective on peace and conflict studies would seem to promise more meaningful results, especially when we aim to understand local agency. Including a spatial perspective in that sense means engaging seriously with the concepts behind the notion of 'the local' (which is a spatial term in itself) and linking agency to the spaces in which it unfolds.

Notes

1. For a more detailed analysis of the mutual entanglement of the global and the local, see Buckley-Zistel and Henrizi (2013).
2. All Iraqi NGOs except for Al Amal are women's NGOs. Al Amal has a strong focus on gender and is part of the Iraqi women's network but it does not focus on women exclusively.
3. This chapter and the observation of the CEDAW process is part of a PhD project that allowed me to work with local NGOs, beginning in 2012.
4. Pursuant to the implementation of UN Security Council Resolution 1325, the Security Council calls on member states to develop NAPs or other strategies

at the national level. As of February 2014 the Iraqi Government announced a NAP, but the draft – if it existed at all – was not accessible to NGOs.
5. For a more detailed impression of the hall, see the CEDAW 57 webcast: http://www.treatybodywebcast.org/cedaw-57-session-iraq-arabic-audio/
6. The Jaafari law is a draft law on civil rights that would deeply restrict women's rights and freedom, subordinating them to conservative religious interpretations of the Quran. For example, it considers nine-year-old girls to be eligible for marriage.

References

Barassi V (2013) Ethnographic Cartographies: Social Movements, Alternative Media and the Spaces of Networks, Social Movement Studies. *Journal of Social, Cultural and Political Protest* 12(1): 48–62.

Berking H (2006) Raumtheoretische Paradoxien im Globalisierungsdiskurs. In: Berking H (ed.) *Die Macht des Lokalen in einer Welt ohne Grenzen.* Frankfurt: Campus, pp. 7–24.

Buckley-Zistel S and Henrizi A (2013) Relating Spaces. The Global, the Local and the Building of Peace. IAPCS Conference Power and Peacebuilding, Manchester, UK, 12–13 September 2013.

Caglar G, Prügl E and Zwingel S (eds) (2013) *Feminist Strategies in International Governance.* Abingdon: Routledge.

CEDAW (2014) CEDAW 57th Session: Iraq – Arabic Audio. Available at: http://www.treatybodywebcast.org/cedaw-57-session-iraq-arabic-audio/ (accessed 28 October 2014).

Get Involved (2013) Call for Local to Global Programme. Available at: http:// www.awid.org/Get-Involved/Calls-for-Participation2/Call-for-Local-to-Global -Program-NGO-Preparation-for-CEDAW-Advocacy (accessed 25 October 2014).

Helfferich C (2012) Einleitung: Von roten Heringen, Gräben und Brücken. Versuche einer Kartierung von Agency-Konzepten. In: Bethmann S, Helfferich C, Hoffmann H and Niermann D (eds) *Agency: Qualitative Rekonstruktionen und gesellschaftstheoretische Bezüge von Handlungsmächtigkeit.* Weinheim and Basel: Beltz/Juventa, pp. 9–38.

Henrizi A (2013): CEDAW Shadow Report Training Beirut 9–11 May 2013. Minutes from the shadow report training in Beirut. Unpublished.

Henrizi A (2015) Building Peace in Hybrid Spaces: Women's Agency in Iraqi NGOs. *Peacebuilding* 3(1): 75–89.

Iraqi State Report (2013): Combined Fourth, Fifth and Sixth Periodic Reports of States Parties: Iraq. Available at: http://tbinternet.ohchr.org/_layouts/ treatybodyexternal/Download.aspx?symbolno=CEDAW%2fC%2fIRQ%2f4-6& Lang=en (accessed 23 January 2015).

IWRAW (2015) About Us. Available at: http://www.iwraw-ap.org/organisation/ about-us (accessed 1 January 2015).

IWRAW (2010a) NGO Participation. Available at: http://www.iwraw-ap.org/ resources/pdf/NGO_Participation_in_CEDAW_Part_1_and_2_Feb_2010.pdf (accessed 25 October 2014).

IWRAW (2010b) Global to Local. Available at: http://www.iwraw-ap.org/ programmes/globaltolocal.htm (accessed 19 September 2014).

Löw M (2001) *Raumsoziologie*. Frankfurt: Suhrkamp.

Löw M (2008) The Constitution of Space: The Structuration of Spaces through the Simultaneity of Effect and Perception. *European Journal of Social Theory* 11(1): 25–49.

Mac Ginty R (2011) *International Peacebuilding and Local Resistance. Hybrid Forms of Peace*. Basingstoke: Palgrave.

Massey D (2004) Geographies of Responsibility. *Geografiska Annaler, Series B, Human Geography* 86(1): 5–18.

Massey D (1994) *Space, Place and Gender*. Cambridge: Polity Press.

OHCHR (Office of the High Commissioner for Human Rights) (2014) Committee on the Elimination of Discrimination against Women: 57th Session, February 2014. Available at: http://www2.ohchr.org/english/bodies/cedaw/cedaws57.htm (accessed 25 October 2014).

OHCHR (2015) Committee on the Elimination of Discrimination against Women: Introduction. Available at: http://www.ohchr.org/EN/HRBodies/CEDAW/Pages/Introduction.aspx (accessed 25 October 2015).

Schlichte K and Veit A (2007) Coupled Arenas. Why State-Building Is so Difficult. *Working Papers Micropolitics* No. 3, Humboldt University Berlin, Germany. Available at http://www.researchgate.net/profile/Klaus_Schlichte/publication/252429338_Coupled_Arenas_Why_state-building_is_so_difficult/links/0f3175 3c78e21367be000000.pdf (accessed 12 June 2015).

Shadow Report (NGO Coalition of CEDAW) (2014) Iraqi Women in Armed Conflict and Post Conflict Situation. Shadow Report submitted to the CEDAW Committee at the 57th Session, February. Available at: http://tbinternet.ohchr.org/Treaties/CEDAW/Shared%20Documents/IRQ/INT_CEDAW_NGO_IRQ_16192_E.pdf (accessed 12 June 2015).

Ruhne R (2011) *Raum Macht Geschlecht. Zur Soziologie eines Wirkungsgefüges am Beispiel von (Un)Sicherheiten im öffentlichen Raum*. Wiesbaden: VS Verlag für Sozialwissenschaften.

UN (2014) Convention on the Elimination of All Forms of Discrimination against Women: Concluding Observations on the Combined Fourth to Sixth Periodic Reports of Iraq. United Nations Report no. CEDAW/C/IRQ/CO/4–6, 10 March. Available at: http://tbinternet.ohchr.org/_layouts/treatybodyexternal/Download.aspx?symbolno=CEDAW%2fC%2fIRQ%2fCO%2f4-6&Lang=en (accessed 25 October 2014).

UN Women (2009) Convention on the Elimination of All Forms of Discrimination Against Women. Available at: http://www.un.org/womenwatch/daw/cedaw/ (accessed 25 September 2014).

UN Women (2007) NGO Participation at CEDAW Sessions. Available at: http://www.un.org/womenwatch/daw/ngo/cedawngo (accessed 17 October 2014).

Zwingel S (2005) From Intergovernmental Negotiations to (Sub)national Change: A Transnational Perspective on the Impact of CEDAW. *International Feminist Journal of Politics* 7(3): 400–424.

Zwingel S (2012) How Do Norms Travel? Theorizing International Women's Rights in Transnational Perspective. *International Studies Quarterly* 56: 115–129.

Part III

Boundaries and Borders

7
Space, Class and Peace: Spatial Governmentality in Post-War and Post-Socialist Bosnia and Herzegovina

Elena B. Stavrevska

Introduction

The hegemonic framework used for both policy-making and analysis not only of the war in BiH[1] but also of the post-war period remains one driven by the logic of 'groupism'. The term, coined by Brubaker (2002), refers to the tendency to approach groups, including ethnic groups, as the basic units of analysis and the main protagonists in social life, with interests and agency as their attributes. This underlying logic was reflected in the General Framework Agreement, also known as the Dayton Agreement, with which the Bosnian War was concluded in November 1995. As part of the agreement, the conflict parties also adopted the Constitution of BiH (Annex 4 of the Dayton Agreement), which postulates the three ethnic groups – Bosniacs, Serbs and Croats – as the sovereign power-holders in the country. Such a model of governance is based on the assumption that the ethnic groups are homogenous, fixed and mappable.

This has set the basis for ethnicity to become the single most important category in Bosnian society. Ethnicity is at present the dominant centre of power, superior even to the state. As a result, we witness a process of 'spatialization of ethnicity' – that is, the imagining of ethnicity as a category that possesses spatial characteristics. It is imagined as higher and wider than any other category in the society, including civil society, local community and family. The spatialization of ethnicity, along with the assumption of ethnicity being mappable, is the foundation of ethnic 'spatial governmentality'. In other words, spatial governmentality refers to the mechanisms used in governing spaces, which is a tendency to separate and isolate, rather than punish certain behaviours and people. These mechanisms combine state practices, including bureaucratic routines and regulations, and individual self-governance. This chapter provides examples of practices and metaphors used in the mutually constitutive process of these two tendencies – the spatialization of ethnicity and spatial governmentality – resulting in ethnic spaces.

The ethnic spaces are territorial and mental, imagined. They are what Harvey (2006) calls 'relational spaces'. These are spaces that are constituted and populated by the way people experience them. In other words, the relationality is closely linked to people's identity. The everyday is the realm within which these spaces are (re)produced. Constituting a set of experiences, practices and interpretations, but also of views, values and interests, aside from the hegemonic ethnoterritorial narrative, everyday experiences are also informed by needs and socioeconomic concerns. Interestingly, while examples of the existence of the ethnic spaces are very much present in the public discourse in BiH, little is spoken of the spaces that are created based on socioeconomic status, both within and across the ethnic spaces.

Most of the research on the interplay between socioeconomic inequality and conflict appears to show a positive linear relationship between the two. However, the role that class structures play in the aftermath of a war has been less examined. This might be due to the dominant assumption that the war has wiped clean the historical structures of class, various inequalities, mentalities and priorities. On the contrary, evidence shows that conflicts in fact accentuate class differentiation. Aiming to explore this issue further, this chapter engages with the question of what role class structure can play in the creation of spaces of peace and solidarity in divided societies.

Through the analysis of different intersubjectivities, notably those of people living with and beside each other, the presented Bosnian examples provide an ethnographic perspective on human sociality and predicaments of everyday life. Drawing on these examples, the main argument is that class structure, a life dimension thus far perceived as fuelling conflict, can in fact create experiential conditions that are needed for spaces of peace to be produced and for ethnic divisions and conflict to be overcome.

This chapter is based on ethnographic fieldwork within a cumulative period of nearly nine months spent in various parts of BiH during 2009–2012 at several locations, including Sarajevo, East Sarajevo, Tuzla, Bijeljina, Gračanica, Doboj, Teslić, Drvar, Banja Luka, Jajce, Brčko and

villages near these towns. The methods used included balanced participant observation and semistructured interviews with informants who
both have experience in and are knowledgeable about the topic. Using
various entry points and gatekeepers, the selection process included
snowball sampling until a certain point of data saturation was reached.
In addition, having been embedded in Western academia, but being
from another former Yugoslav country and speaking the local lan-
guage(s), has provided me with an insider/outsider position, which has
informed my biases as a researcher. With the locals perceiving me as
'theirs' (*naša*) due to my positionality, I obtained access to valuable
insights into people's everyday experiences and into the topic of this
chapter. Despite the inevitability of being more part of some networks
than of others, I aimed to hear as many diverse voices as possible by
gaining access to different gatekeepers and actively pursuing informants
from outside the familiar networks.

In the analysis of space, this chapter draws on social geographers
who have built on Marxian spatial approaches, on the one hand, and
on anthropologists who have long been ethnographically describing
the spatial consequences of state practices, on the other. Coming from
critical peace studies, this has allowed me to analyse the post-conflict
realities outside the 'groupism' logic, which is almost inherent in peace
and conflict studies. To that end, a spatial perspective adds a valuable
component to the analysis of post-conflict societies by focusing on the
spaces that people produce, occupy and interpret, rather than on the
groups with which they self-identify.

This chapter proceeds by first elaborating on the phenomena of
spatialization of ethnicity and spatial governmentality, and the interaction thereof, and follows with an examination of the ethnic spaces
that are produced through that interaction, and examples of metaphors
and practices through which they are reinforced. By zeroing in on the
everyday, it then sheds light on the importance of class structure and
socioeconomic commonalities in overcoming the divisions and producing spaces of peace and solidarity. Finally, the chapter concludes with
a number of examples of intersubjectivities whereby class has played a
positive role in setting the basis for reconciliation.

Spatialization of ethnicity and spatial governmentality

Scholars from different disciplines, in particular anthropologists, have
long been examining the imagination and production of the state
through social practices. An important component of that analysis has

focused on the ways in which states are spatialized and people get to experience them as entities with spatial characteristics. Exploring the relationship between state, space and scale, Ferguson and Gupta (2002) shed light on a process that they refer to as 'the spatialization of the state'. This is the operation of metaphors and practices through which 'states represent themselves as reified entities with particular spatial properties', which contributes to them '[securing] their legitimacy, [naturalising] their authority, and [representing] themselves as superior to, and encompassing of, other institutions and centres of power' (Ferguson and Gupta 2002: 981–982). The spatialization of the state relies on two key principles: 'verticality', which is the idea that the state is higher or above the rest of the society, reflected in the common description of state decisions as 'top-down', and 'encompassment', which is the idea that the state is wider and larger than anything else in the society, including family, community and civil society (Demmers and Venhovens: Chapter 8; Ferguson and Gupta 2002: 982). These are embedded in the routinized state bureaucratic practices, through which spatial orders are reproduced.

However, in instances when the state is hollowed out, other categories become politically more salient and are spatialized in a similar manner. A particularly fertile type of states for such processes, for example, are post-conflict societies with a consociational form of government, whereby the state is seen primarily as the negative image of the conflict and a means of managing conflicts between the different, often ethnic, communities in the country. One such case is BiH and the Dayton Agreement, Annex IV of which is the Constitution of BiH, which in addition to successfully ending the war also postulated ethnicity as the sole most salient cleavage in the Bosnian political system. As a result, the whole political structure of Bosnia is based on the same identities around which the conflict was constructed. Some 20 years after the war ended, politics in the country remains centred on the issues of ethnicity, political representation and collective political rights of the three nationalities, with ethnoterritorial politics dominating the public discourse. Therefore what we witness in BiH is the spatialization of ethnicity, not the state, as the main centre of power.

In terms of verticality, on the one hand, post-war Bosnian society appears to assume a certain hierarchy whereby ethnicity is the highest category and consequently has the greatest claim to universal significance. According to the constitution, the country has three constituent peoples (Bosniacs, Serbs and Croats), which share the political power. This leaves Bosnian society to be governed not as a sovereign political community, *demos*, but as a conglomerate of three different ethnic communities, *ethos* (Hayden 2005: 226). In that sense it is ethnicity that is seen as the dominant centre of power, which is reflected in countless practices and metaphors.

When it comes to encompassment, on the other hand, as a consequence of the dissolution of Yugoslavia and the Yugoslav Wars resulting both in large overseas diasporas and in ethnic communities now being divided with parts of them residing in neighbouring countries, ethnicity is perceived as a wider category than any other. This is reflected in everyday references to common history, language, religion and myths of a certain ethnic group.

The operation of these two overarching ideas makes it possible for the people of Bosnia to experience ethnicity as a category with spatial characteristics. Moreover, they are embedded in the bureaucratic practices of the Bosnian state at all levels of government through which ethnicity becomes further fused with territory and particular territorial spaces. The constitution, for instance, specifies that the country consists of two entities – the Federation of BiH (FBiH) and the Republika Srpska (RS) – the former of which is later in the text related to the Bosniacs and the Croats, and the latter with the Serbs in the country. A few years ago, as a result of an initiative by the international community in BiH, the constitutions of the FBiH and the RS were amended to acknowledge all three constituent peoples within each entity (Constitution of the Federation of Bosnia and Herzegovina 1994; Constitution of Republika Srpska 1992). Article 1 of the Constitution of RS, for instance, now states that 'The Serbs, Bosniacs, Croats, as constituent peoples, Others and citizens shall participate in executing the functions of authority in the Republic equally and without discrimination' (Constitution of Republika Srpska 1992, Article I, Section 1, Clause 4). Despite that, however, there is a silent understanding among the people about which municipalities, towns and villages 'belong' to which ethnicity. To that end, ethnicity in Bosnia is to a great extent perceived as a mappable category, and the spatial consequences of the ethnoterritorial nature of politics in BiH are ethnically conceived territorial spaces.

Such spaces are produced and reproduced through mechanisms that are a form of spatial governmentality. These mechanisms used for social ordering based on the regulation of spaces and as a form of regulation depend on the creation of spaces that are characterized by the consensual, participatory governance of selves (Merry 2001: 20). In other words, the focus is on managing the spaces people occupy rather than managing the people themselves; the target is a population, not individuals. In that sense, the aim is not to reform certain behaviours but rather to isolate and separate socially undesirable behaviour. Instead of punishing 'offenders' in an open space, regulations and various security practices make it impossible for potential misfits to enter the 'closed' space.

In addition to the special ordering carried out through practices and regulations, spatial governmentality also involves individual self-governance. It is in fact dependent on the 'neoliberal regime of individual responsibility and accountability' (ibid.: 17). These spaces are thus relational spaces. The relational concept of space, developed by Harvey (2006: 273), views space as intimately linked to the processes that define it, with the very concept of space being embedded in or internal to the process. Interested in the implications of relational spaces to political subjectivities, Harvey (ibid.: 277) argues that 'what we do as well as what we understand is integrally dependent upon the primary spatio-temporal frame within which we situate ourselves.' To that end, while related to various territorial spaces, spatial governmentality is dependent on relationalities. With ethnically conceived spaces, regulations and bureaucratic practices do contribute to the ongoing creation of social orders and identifications that occur as a result of the governing of those spaces (Hromadžić 2011: 271), but the identification itself is just as important in the governing process.

Governmentality of ethnically conceived spaces

One manifestation of governmentality of ethnically conceived spaces in BiH can be found in the schools, both primary and high schools. In other words, Bosnia is widely known for the 'two schools under one roof' (*dve škole pod jednim krovom*) phenomenon (BBC 14 September 2012). This is a system where two schools with different management and following different teaching curricula function in the same building, with the children of the different schools and therefore different ethnicities attending school in different shifts, or having separate entrances or separate classrooms. This phenomenon is most common in the Central Bosnia and the Herzegovina-Neretva cantons, where the majority of the students are Bosniacs and Croats, studying separately (Hrnjić 2012, personal interview).[2]

Certain separation among the students exists even in places that do not practice this system. That is, even though most of the curriculum content used in the elementary schools around the country has been jointly agreed on to be universally used and was signed into law by the

Common Core Curriculum in 2003, there remains a so-called 'national group of subjects' which in most places has included mother tongue, religious instruction, history, geography, nature and society/my environment, and music and art (in the case of the Brčko district) that students attend separately (OSCE BiH 2009). While the situation since 2010 has improved in most places, the subjects of mother tongue and religious instruction almost universally remain ethnicity/religion based. It should also be noted that while in most schools in the FBiH and the Brčko district there is the option of attending an alternative areligious course (usually history of religion, life skills or culture of religion), most students still opt to attend religious instruction, mainly each ethnicity studying 'its religion'. Such an alternative is not offered in RS (Kojić and Idžan 2012, personal interviews).[3] Similarly, places in RS where attending and learning a mother tongue other than Serbian 'is required' are rare. This example shows that while, on the one hand, the education authorities themselves limit the spaces to certain ethnicities only by offering certain courses, the people too participate in the process by complying with the governing expectations and not enrolling their children in schools and courses seen as intended for other ethnicities.

Another manifestation can be observed through the use of alphabets. The Cyrillic alphabet inside BiH appears to be perceived as intimately linked to the Serbian people and the Serbian language, with signposts in RS being written first in Cyrillic then in the Latin alphabet, or in some cases exclusively in Cyrillic, as, for instance, the names of the streets in Bijeljina, and vice versa in the FBiH, as in the case of Tuzla. Similarly, as evidence of protest, one often sees the Cyrillic or the Latin place names, depending on where one finds oneself, being crossed out with spray on signposts. There have also been instances where university professors in Banja Luka would refuse to grade or even accept a written exam that was not written in Cyrillic (University of Banja Luka alumni 2012, personal interviews).[4]

While both alphabets are constitutionally declared as the official scripts of the country, and by third grade each child is required to have learnt both, the reality is quite different. Some teachers push for the learning of both alphabets, among other reasons due to the availability of library books in 'the other script' (Tuzla primary schoolteachers 2012, personal interviews),[5] yet there are also teachers who are not familiar with the 'other' alphabet themselves and are unable to teach it to their students (international organization representative 2012, personal interview).[6] A result of these circumstances is incidents like the case when students from Livno (FBiH) were to travel to Mrkonjić Grad (RS) for a school competition and were unable to communicate their exact location to the organizers due to the signposts being written in a script that was unreadable to them: Cyrillic (international organization representative 2012, personal interview).[7] The spatial governmentality of the linguistic component of ethnic spaces is embedded in the legal framework in BiH. For instance, Article 6 of the FBiH Constitution states that '[t]he official languages of the Federation shall be the Bosniac languages and the Croatian language. The official script will be the Latin alphabet' (Constitution of the Federation of Bosnia and Herzegovina 1994, Article I, Section 1, Clause 1). The majority of people comply with such a vision of linguistic and alphabetic division.

More mundane and symbolic mechanisms of spatial governmentality involve religious practices and flag display. Each ethnicity in Bosnia has been closely identified with a certain religion, with the majority of Bosniacs perceived as being Muslims,[8] the Serbs as Christian Orthodox and the Croats as Christian Catholics. In that sense, religion plays an important role in demarcating the different ethnic spaces. Every school, as well as various institutions in RS, for instance, celebrate their patron saint day, with many of them having an Orthodox priest present on the day for the ritual of bread blessing.[9] There have also been examples of Catholic priests leading a prayer and blessing bread in Croatian elementary schools in BiH at the school celebration of Bread Day. In addition, in RS the school choirs sometimes prepare and sing at celebrations the anthem of Serbia, *Bože pravde*, the text of which centres on the 'Serbian people and the Serbian lands', which has caused the non-Serbs to leave the choirs.[10]

Another mundane mechanism is the display of flags. Seeing the BiH state flag in RS is not very common, for example, while the RS flag, and even sometimes the flag of Serbia, can be seen displayed in most places. Similarly, the Croatian flag and even the Croatian coat of arms can be seen all over Mostar, Livno, Jajce and other places where there is a larger Croatian population. All across the FBiH one also notices the so-called 'Bosniac flag with golden lilies', but also various Islamic flags, with the green-with-white-crescent-and-a-star flag being displayed on every mosque. This (re)production of ethnically conceived spaces through flag display operates through both institutional and individual practices. The individual spatial self-governance is evident even in less dramatic social routines. By way of illustration, in multiethnic places, such as Brčko or Livno, there is quiet agreement about which ethnicity goes to which bars. While no one should be refused entrance to a bar where they would be a minority, people rarely cross that line.

The role of class: Beyond ethnic spaces?

The existence of exclusionary ethnic spaces, which as relational spaces are both territorial and mental categories, leaves the citizens by and large living and functioning in separate ethnic spaces, with very few opportunities for interethnic or supraethnic spaces to emerge. Importantly, in a conflict that is presented through the dominant discourse as an ethnic one, supraethnic spaces also constitute spaces for peace and grassroots reconciliation.

The examples above highlight the role of state practices in (re)producing ethnic spaces, and through them ethnic divisions. They also shed light on the essential role of the everyday and the interconnectedness thereof with the practices of the state. In that sense the everyday is about the socially and 'culturally appropriate form of individual or community life and care' (Richmond 2011: 15). In addition to the social, the everyday is by and large also a realm of the political. Adopting this understanding is an attempt to step away from the Western construct of the political, mindful of the case-study context and the 'growing cynicism about what can be accomplished through the practices and institutions of politics [which] has led to an upsurge of interest in exploring other sites of change – as in culture and everyday life – in the belief that politics, as traditionally understood, can be redirected from outside' (Darby 2006: 48).

Constituting a 'set of experiences, practices and interpretations through which people engage with the daily challenges of occupying, preserving, altering and sustaining the plural worlds that they occupy' (Mitchell 2011: 1624), the everyday is populated with views, values and interests, which aside from the dominant narrative shaped around the issue of ethnicity are also informed by everyday needs. To that end, the everyday experiences could also set the basis for overcoming ethnic divisions when grounded in the commonality of needs and socioeconomic interests.

The link between socioeconomic inequality and conflict or political violence has long been studied, with most scholars agreeing that there is a positive linear relationship between the two (Muller 1997; Nafziger and Auvinen 2002; Russett 1964; Zimmerman 1980). That is, there is a scholarly consensus that inequality produces conflict. At the same time, little attention has been paid to what role historical structures of class can play in the production of spaces for peace, if not between then within classes. The vast majority of the post-war economic state-building and peacebuilding efforts since the end of the Cold War, informed by neoliberal logic, have focused on creating capable governing institutions and liberalizing markets. However, this is based on the false premise that the war wiped everything clean, including various inequalities, mentalities, traditions and priorities from before, and what was left was a *tabula rasa* which can be filled in accordance with the current agenda (Gilbert 2006: 17). On the contrary, evidence suggests that the war in actuality accelerates the ongoing processes of socioeconomic differentiation and class formation (Cramer 2008). In the case of Bosnia, the remains of the previous system seem to be part and parcel of today's Bosnian society and are crucial in the shaping of the lived experiences, knowledge and sensibilities of the people who continue to dominate all spheres of life in Bosnian society (Kurtović 2010).

Yet, with most of the conflicts in recent years being conceptualized as essentially cultural, the hegemonic discursive framework has not been taking into account the historical structures of class. Rather, its underlying logic has been that of 'groupism'. This concept, coined by Brubaker (2002), captures the tendency to ascribe agency characteristics and interests to groups, in this case ethnic groups, and to treat them as the core protagonists of social life. This logic has been employed by scholars and policy-makers alike, and it has become embedded in and extended through various routinized practices in society, resulting in the abovementioned spatialization of ethnicity. As a consequence of the dominance of this discourse, the logic becomes fundamental in how peacebuilding practitioners, scholars, civil society organizations, politicians and citizens imagine and inhabit a particular post-conflict society.

Importantly, both such logic of groupism and the spatialization of ethnicity are based on the assumption that ethnicity is a higher category and has a more prominent role in people's lives than any other. The mechanisms through which the spatial governmentality of ethnic spaces operates also assume certain equality within the spaces – that is, within the ethnicity that occupies the particular space. More fundamentally, the spatial governmentality of ethnic spaces assumes precedence of the commonality that ethnicity provides over any potential differences and inequalities, including socioeconomic ones. In reality, however, even though insignificantly present in the public discourse, class spaces both within and across ethnic lines exist. For instance, the Poljine residential settlement, some 20 minutes from the centre of Sarajevo, is known as the place where the Sarajevo elite reside. At the same time, the communist-era residential apartment buildings that survived the war are still mainly inhabited by working-class families.

Spaces of peace and solidarity in BiH

Class structure, as a dimension of life that is often seen as fuelling conflict, can in fact also create experiential conditions of possibility to overcome ethnic divisions and conflict. This can be examined by moving the analytical perspective beyond the logic of groupism and ethnic spaces, and focusing on the everyday socioeconomic predicaments and lived experiences in a post-war society and how they inform intersubjectivity.

Drawing on extended ethnographic work, this chapter examines two types of intersubjectivity. Phenomenologist Schutz (1967) distinguishes between four different types of spatial and temporal intersubjectivity: predecessors, successors, contemporaries and consociates. The latter two, in reference to people living at the same time, are the two of interest here. Contemporaries are understood to be people who exist at the same time and live next to each other. Carrithers (2008: 167) describes them as people who 'form a world with which we live, but not through intimate involvement and the discovery of individual life stories and vicissitudes, but only through our knowledge of their social genus, and the accompanying knowledge of their appropriate roles, and of our own roles in respect to them'. Consociates, on the other hand, are people who inhabit the same time and space, and live with each other. In the words of Carrithers (2008: 167),

> Consociates are people we grow old with, whose lives we participate in, whom we know intimately and in their own terms. We are entwined with them; we are able to join in their absolutely individual life story, and to that extent, we see beyond any generic designation to particularities of attitude, experience, and reaction. We have, with consociates, a 'thou-relationship', an intimacy and mutual knowledge of one another face to face, and a 'we-relationship', in that we have experiences in common with them.

In analysing the interaction of contemporaries and the potential that class and socioeconomic commonalities have in creating the nexus of a supraethnic space, an important example can be found in the February 2014 social protests in BiH. On 5 February 2014, in the city of Tuzla, workers from companies that were privatized and went bankrupt, supported by students and activists, took to the streets to protest against the failed privatization, the dysfunctional system, the high unemployment rate and the corrupt ethnonationalist political elites (Eminagić 2014: 1).

The initially peaceful protests, organized by the workers' unions, ultimately escalated into violence. In the next few days the protests spread across the country, from Sarajevo to Zenica, Mostar, Brčko, as well as Banja Luka, albeit on a smaller scale. Being the biggest protests since the war, they resulted in several resignations, including those of cantonal governments and prime ministers.

More importantly, these were the first large-scale social protests that brought people together using a different narrative from the one that had dominated the everyday life of Bosnian and Herzegovinians in the post-Dayton period. That is, the protesters were mainly working-class people. What followed was the creation of citizens' plenums in several cities, as a method of direct democracy, where the assemblies, open to all citizens, discussed and formulated clear demands from the governing structures. The plenums, which were active in the months after the protests, in the form of both physical spaces where the assemblies took place and virtual or mental spaces populated by those sharing the same socioeconomic predicament, appeared to become supraethnic spaces, where people united based on their class, rather than their ethnic background. Gathered under the 'We are hungry in three languages' and 'End nationalism' slogans, in the words of one of the most prominent members of the Tuzla plenum, 'people who had never met before, and who had hitherto lived in their separate social circles, came together, jointly, to make demands, and in so doing, transformed public space into social space' (Arsenijević 2015: 8). Founded in the shared discourse, a form of commonality emerged. This is a case not only of a narrative that challenges the divisions and promotes solidarity but also a possibility for development of a space where genuine reconciliation can take place.

As far as consociates are concerned, examples of resistance (even if unintended) to the ethnic spaces, which provides the basis for a space that possibly enables grassroots reconciliation and solidarity to emerge, can be found in ethnically mixed rural areas where living together and helping each other are a matter of survival. While these locations are few, they are frequently far from the urban centres and are predominantly populated by elderly people. Even though many of their inhabitants remark that a sufficient amount of time after the war had to pass for them to (re)build relationships and trust, they now not only cooperate out of a rational interest but are also friends. Fata from a village near Tuzla explains how she shares seeds with her Serbian neighbour, Anica, and how their facing similar challenges by living in the same village has brought them closer together. Similar examples can be found along the Inter-Entity Boundary Line, either in the villages that the line runs through or those in its immediate proximity. Instances like these are pockets of in-between ethnic spaces cohabitation, which leaves room for reconciliation to eventually occur. This is where the phenomenon of *komšiluk* (neighbourhood) comes into play. *Komšiluk* implies a certain level of closeness and community, a group of consociates. It 'consists of bonds, relations and imagination that are cultivated in the flows of everyday sociality' (Henig 2012: 10). In the realm of *komšiluk*, especially in rural areas, it is common for neighbours to assist each other with different work and activities, but also to have a strong personal bond. This appears to be the case even in those villages and neighbourhoods with an ethnically heterogeneous composition. The traumatic experiences of the war, needless to say, have also been a factor in the creation of this in-between, supraethnic space, as well as the pre-war experiences of the people. To that end, also taking into consideration the large migration of younger people to the country's urban centres, such supraethnic spaces are primarily created among elderly people, or at the very least people who lived together prior to the war and have a certain socioeconomic commonality, reflected in common values and interests. The young people, on the other hand, remain by and large within the ethnic spaces. Thus most of those living in these pocketed spaces of peace and solidarity are retired working-class people.

While, as Stefansson rightfully notices, working or having coffee together does not necessarily symbolize deep love or friendship, it is nonetheless an expression of respect and an acknowledgement of a certain commonality (Stefansson 2010: 68–69), which in some cases is a shared socioeconomic status. Such is the example of the taxi drivers. In many places, such as Tuzla, Srebrenik, Brčko, Doboj and Jajce, there are fixed prices for taxi rides, which in some cases have been established by the local municipality, while in others through an agreement of the association of taxi drivers. What is interesting is their approach towards such fixed prices, with many of them saying that it is 'enough', or that everyone's standards are low these days, or that they are not going to get richer or poorer because of those few Bosnian convertible marks.

Another curious example is the silent agreement that exists among the taxi drivers in some places, such as Brčko, over who covers a particular spot (e.g. bus station, hospital, city centre), even though they each work for themselves (Brčko taxi drivers 2012, personal interviews).[11] There is also cooperation among them, with some drivers passing on customers to others if they are not in that area at the moment of the call, even if the customer agrees to wait. 'Let him earn something today too. I have already made enough,' says one of them (Brčko taxi driver 2012, personal interview).[12] The cooperation in Doboj is arranged by the local municipality, with it having divided the taxi drivers into four groups of six drivers, each group covering a different location every day. 'Those that are at the train station today will earn less, but tomorrow when we rotate they will be at the bus station and they will compensate for the previous bad day,' explains a Doboj taxi driver. 'We have to cooperate so that there is something for all of us, even if each of us is registered as a separate firm' (Doboj taxi driver 2012, personal interview).[13] Depending on the size of the town, many of them point out that they get along well with each other, often even having coffee together.

The basis that shared socioeconomic concerns provide in overcoming the animosities from the war and reconciling is perhaps best illustrated by the statement of an interviewee from Drvar, a predominantly Serbian town located in a predominantly Croatian canton in the FBiH, who said: 'we no longer care who is Serbian and who is Croatian, we will all likely have to leave the town because of unemployment anyway. We are united in our struggle to ensure subsistence. United in our tragedy' (Drvar citizen 2012, personal interview).[14]

Conclusion

With most of the conflicts in the post-Cold War era being by and large conceptualized as cultural to the very core, the hegemonic analytical framework has been one characterized by 'groupism', whereby ethnic groups are conceived as entities and examined as actors in their own right. Policy-makers, scholars, journalists, peacebuilding practitioners and citizens alike continue to approach ethnicity, race and nation as fixed, internally homogenous and often mappable categories. The latter relates to their spatialization.

What we witness in BiH is a case of spatialization of ethnicity, with ethnicity being the most powerful political category, superior to the state. In fact, the state is imagined merely as a conglomerate of the three ethnicities in the constitution referred to as 'constitutive peoples' – the Bosniacs, the Serbs and the Croats. Consequently, ethnicity appears to have the greatest claim to universal value and significance in the country and is imagined as having both verticality and encompassment, resulting in people ascribing certain spatial characteristics to it. This spatialization of ethnicity has been widely accepted by the governed and has been the basis of ethnic spatial governmentality. The spatialization is crucial in how, *inter alia*, policy-makers, scholars and citizens imagine and inhabit post-war BiH, resulting in a constant reinforcement of ethnically conceived spaces. These spaces are relational, meaning that while being territorial in nature they are in a mutually constitutive relationship with the processes that define them. The relationalities provide an ordering system for both individual and collective thoughts, as well as perceptions and feelings. This is fundamental for the mechanisms of spatial governmentality, through which spaces, not people, and populations, not individuals, are managed. Spatial governmentality is dependent both on bureaucratic practices and regulations and on individual self-governance. These two tendencies – the spatialization of ethnicity and spatial governmentality – work in parallel and often complement each other, resulting in ethnically conceived spaces. This chapter provides examples from Bosnia of the mutual constitution of spatialization and governmentality through metaphors and practices.

On the other hand, in examining conflict societies, there has been a lot of research that shows a positive linear relationship between socioeconomic inequality and conflict. Nonetheless, the role that socioeconomic issues, in particular historical structures of class and socioeconomic commonalities, play in the post-war period has not received much scholarly attention. This might be particularly so because the dominant approach in conflict analysis has been one based on the assumption that the war has wiped everything clean, including class differentiation. Not only is this a false assumption but – on the contrary – evidence suggests that conflict accelerates class formation and social strata differentiation. This chapter addresses the question of what role class structure can play in the creation of spaces of peace and solidarity.

Examining the available ethnographic evidence, the chapter employs Schutz's distinction between contemporaries and consociates as types of intersubjectivity whereby the former are people who live next to each other, recognizing each other's social genus, while the latter are people living with each other – ones that have a 'we relationship'. In the case of the first type of intersubjectivity, using the example of the February 2014 social protests in several towns in BiH, in the aftermath of which citizens' plenums were formed, resulting in the emergence of a certain sharedness and commonality based on class struggles, they showed the potential that class has in serving as the basis for the development of supraethnic spaces, or spaces of peace. As for the second type, the consociates, everyday predicaments were closely analysed through the examples of *komšiluk* and people living in rural areas near or next to the Inter-Entity Boundary Line, which is where governance boundaries overlap and/or unintentionally leave an 'undergoverned grey area' in between, where grassroots actors can fully exercise their agency. Equally importantly, these are also areas where people often struggle for subsistence. Another presented example of the class–space–peace interplay relates to the taxi drivers in towns where they have to self-manage the sector in order to ensure subsistence for all of them. These are all spaces where the existing governmentality of ethnic spaces is challenged and reconciliation can or does occur. This chapter thus argues that these processes and issues of class can, in fact, contribute to the creation of spaces of peace – that is, spaces where reconciliation can occur at the grassroots level. Through everyday experiences, this dimension of life that is often seen as fuelling conflict – class structure – can create experiential conditions of possibility to overcome ethnic divisions and conflict.

While more research on the forms in which socioeconomic commonality and class structures interact with the peace processes at the grassroots level is needed, this chapter sheds light on the necessity of a nuanced approach when analysing spatiality and spaces of conflict or peace in post-conflict societies.

Notes

1. Bosnia and Herzegovina refers to two geographically distinct parts of the country. While acknowledging this fact, in the text I nonetheless refer to the country as BiH and Bosnia interchangeably.
2. Personal interview with N. Hrnjit, former mayor of the municipality of Jajce, Jajce, 27 June 2012.
3. Personal interviews with Danijela Kojić and Ivana Idžan, Ministry of Education and Culture of Republika Srpska representatives, Banja Luka, 23 November 2012.
4. Personal interviews with University of Banja Luka alumni, Banja Luka, 23 July 2012.
5. Personal interviews with primary schoolteachers, Tuzla, 3 and 5 June 2012.
6. Personal interview with an international organization representative, Banja Luka, 26 November 2012.
7. Personal interview with an international organization representative, Banja Luka, 26 November 2012.
8. While Slavic Muslims were called Bosniacs during the Austro-Hungarian rule of the country, in the Yugoslav Constitution it was the *Muslims* who were recognized as a nationality, without having the option to declare themselves Bosniac. The term was reintroduced with the 2nd Bosniac Congress in September 1993 and has since been commonly used, sometimes interchangeably with 'Muslim'. There are two different spellings of the term – Bosniac and Bosniak. Since the former is used in the Dayton Agreement, I use that spelling throughout the text. Importantly, the term 'Bosnian' refers to a citizen of BiH and is not to be confused with 'Bosniac', the ethnic group.

9. Field notes, Teslić and Banja Luka, 20–23 July 2012.
10. Interview with former students, Teslić, 20 July 2012.
11. Personal interviews with taxi drivers, Brčko, 19 and 21 June 2012, 14 November 2012.
12. Personal interview with a taxi driver, Brčko, 21 June 2012.
13. Personal interview with a taxi driver, Doboj, 21 November 2012.
14. Personal interview with a Drvar citizen, Drvar, 24 November 2012.

References

Arsenijević D (2015) Introduction. In: Arsenijević D (ed.) *Unbribable Bosnia and Herzegovina: The Fight for the Commons*. Baden-Baden: Nomos, pp. 7–10.

BBC (British Broadcasting Corporation) (2012) Under One Roof in Bosnia. Available at: http://www.bbc.co.uk/worldclass/19559841 (accessed 19 November 2012).

Brubaker R (2002) Ethnicity without Groups. *Archives Européennes de Sociologie* 43(2): 163–189.

Carrithers M (2008) From Inchoate Pronouns to Proper Nouns: A Theory Fragment with 9/11, Gertrude Stein, and an East German Ethnography. *History and Anthropology* 19(2): 161–186.

Constitution of the Federation of Bosnia and Herzegovina (1994). Available at: http://www.unhcr.org/refworld/docid/3ae6b56e4.html (accessed 24 July 2012).

Constitution of Republika Srpska (1992). Available at: http://www.vijecenarodars .net/materijali/constitution.pdf (accessed 24 July 2012).

Cramer C (2008) From Waging War to Peace Work: Labour and Labour Markets. In: Pugh M, Cooper N and Turner M (eds) *Whose Peace? Critical Perspectives on the Political Economy of Peacebuilding*. London: Palgrave Macmillan, pp. 123–140.

Darby P (2006) Rethinking the Political. In: Darby P (ed.) *Postcolonizing the International: Working to Change the Way We Are*. Honolulu: University of Hawaii Press, pp. 46–72.

Eminagić E (2014) Yours, Mine, Ours? We're All in This Together Now! In: Left East. Available at: http://www.criticatac.ro/lefteast/yours-mine-ours-were-all-in-this-together-now/ (accessed 29 June 2015).

Ferguson J and Gupta A (2002) Spatializing States: Toward an Ethnography of Neoliberal Governmentality. *American Ethnologist* 29(4): 981–1002.

Gilbert A (2006) The Past in Parenthesis: (Non) Post-Socialism in Post-War Bosnia-Herzegovina. *Anthropology Today* 22(4): 14–18.

Harvey D (2006) Space as a Keyword. In: Castree N and Gregory D (eds) *David Harvey: A Critical Reader*. Oxford: Blackwell, pp. 270–293.

Hayden RM (2005) 'Democracy' without a Demos? The Bosnian Constitutional Experiment and the Intentional Construction of Nonfunctioning States. *East European Politics and Societies* 19(2): 226–259.

Henig D (2012) 'Knocking on My Neighbour's Door': On Metamorphoses of Sociality in Rural Bosnia. *Critique of Anthropology* 32(1): 3–19.

Hromadžić A (2011) Bathroom Mixing: Youth Negotiate Democratization in Postconflict Bosnia and Herzegovina. *PoLAR: Political and Legal Anthropology Review* 34(2): 268–289.

Kurtović L (2010) Istorije (bh) budućnosti: Kako misliti jugoslovenski postsocijalizam u Bosni i Herzegovini? *Puls demokratije*. Available at: http://arhiva.pulsd emokratije.net/index.php?&l=bs&id=1979#f38 (accessed 25 October 2014).

Merry SE (2001) Spatial Governmentality and the New Urban Social Order: Controlling Gender Violence through Law. *American Anthropologist* 103(1): 16–29.

Mitchell A (2011) Quality/Control: International Peace Interventions and 'the Everyday'. *Review of International Studies* 37(4): 1623–1645.

Muller E (1997) Economic Determinants of Democracy. In: Midlarsky MI (ed.) *Inequality, Democracy, and Economic Development*. Cambridge: Cambridge University Press, pp. 133–155.

Nafziger EW and Auvinen J (2002) Economic Development, Inequality, War, and State Violence. *World Development* 30(2): 153–163.

OSCE (Organization for Security and Co-operation in Europe) BiH (2009) Primary School Curricula in Bosnia and Herzegovina: A Thematic Review of the National Subjects. Sarajevo: OSCE BiH.

Richmond OP (2011) *A Post-Liberal Peace*. Abingdon: Routledge.

Russett BM (1964) Inequality and Instability: The Relation of Land Tenure to Politics. *World Politics* 16(3): 442–454.

Schutz A (1967) *The Phenomenology of the Social World*. Evanston: Northwestern University Press.

Stefansson AH (2010) Coffee after Cleansing? Co-existence, Co-operation, and Communication in Post-conflict Bosnia and Herzegovina. *Focaal* 57(2): 62–76.

Zimmerman E (1980) Macro-comparative Research on Political Protest. In: Gurr T (ed.) *Handbook of Political Conflict: Theory and Research*. London: Macmillan, pp. 167–237.

8
Bluffing the State: Spatialities of Contested Statehood in the Abkhazian-Georgian Conflict

Jolle Demmers and Mikel Venhovens

Introduction

When travelling through the immense desert of northern Mali in July 2014, one might have run into a small flag, a flag evidently self-made: a piece of hardboard, nailed on a stick and planted in bare land next to a dust road. To the outsider, the small flag might have looked fragile, defenceless, informal, temporary in the middle of the vast, empty space. To those who made and planted it, however, this was the proud evidence of the national flag of Azawad. To supporters of the National Liberation Movement of Azawad (Mouvement National pour la Libération de l'Azawad; MNLA), it marked the beginning of the territory of Azawad. It held authority and demanded respect. On 6 April 2012, the MNLA declared the unilateral independence of the state of Azawad, which consists of two-thirds of the northern part of Mali. Not recognized by any other country in the world, Azawad has been contested and fought over by multiple armed actors, each claiming and performing authority.

By planting the little flag-placard, a border was made. What we see here is social constructivism stretched to its limits. A phenomenon may be imagined, but as soon as people start to act upon an imagined phenomenon, it becomes real in its consequences. A large part of 'bluffing the state of Azawad' is to make it real by mimicking the routines of statehood we see elsewhere in greater sophistication. These are the paraphernalia of statehood: the flag signalling the border, the rubber stamps which are pressed upon the documents of new categories of 'foreigners' and 'citizens', the acting out of security rituals at crossings, and the monuments built to glorify the new state. They all try to discipline us into the scripts and routines of Azawad as an independent state.

Despite observations about how contemporary conflicts have become 'networked', 'post-state' and 'algorithmic', to many parties the state is still seen as the main 'trophy' of war (Amoore 2009; Duffield 2002). Although, to some, shoring up an image of sovereign statehood in the era of global capitalism may seem paradoxical, the fact of the matter is that many warring factions still project notions of self-determination, protection and solidarity onto the state. Claims to 'statehood' make up an important share of contemporary repertoires of contention. But what now is this 'statehood'? How is one to 'think' the state as a form of contentious politics?

This chapter examines contested statehood from a spatial perspective. More concretely, it explores the spatial dimension of the mechanisms of government deployed by parties in the Abkhazian-Georgian conflict to secure their legitimacy, naturalize their authority and represent themselves as superior to rivalling political campaigns. We focus on the making and unmaking of borders – more specifically the border crossing at the Inguri River – as a site of spatial contestation. In line with Merry (2001), we see borders as regimes of governance in which control is exercised through the management of space. Drawing on earlier theoretical work on space, discourse, power (Demmers 2012) and fieldwork in the region (Venhovens 2013), we aim to show how through architectural design, security devices and multiple bureaucratic rituals and routines, authorities from both sides – Georgian and Abkhazian – aim to claim space and produce subjectivities in highly contested ways.[1]

Reading conflict through space: The politics of place, scale and mobility

In the construction of a theoretical approach, this chapter builds on ideas of 'ethnographies of the state'. We aim to contribute to this literature by looking at the spatial dimension of the mechanisms of government deployed by actors in conflict and the ways in which they 'imagine the state'. The literature on state spatiality (Elden 2006; Lie 2002; Mountz 2013; Ó Tuathail 2010) offers a new and distinct research methodology well suited to the analysis of the strategies deployed by conflict actors in their efforts to produce or erase a 'state effect'. The literature on space and spatialization views states not simply as functional bureaucratic apparatuses but as powerful sites of symbolic and cultural production (Joseph and Nugent 1994; Scott 1998). In this context it becomes possible to speak of states, and not just nations as 'imagined' – that is, as 'constructed entities that are conceptualised and made socially effective through particular imaginative and symbolic devices that require study' (Ferguson and Gupta 2002: 981). Through specific sets of images, metaphors and practices, states are represented as concrete and tangible entities with spatial properties. Ferguson and Gupta (2002) refer to the operation of these metaphors and practices as the 'spatializing of the state'. Basically, through mundane and multiple rituals and routines of bureaucratic practice, states produce scalar and spatial hierarchies, expressed by the everyday image of the state as both vertical and encompassing. The notion of verticality refers to imagining the state as somehow situated 'above' society, as hovering above us, as a protective, knowing, god-like force. This is illustrated by the ways in which states are often depicted as pyramids, with the state government at the top and the people at the 'grassroots'. Bodily (and mythical) conceptions of society imagine the state as the rational control centre regulating the passions and appetites of its lower, wilder parts. The second idea is that of encompassment. Here the state is seen as containing a series of circles that begin with the family and ripple out in rings from the local community, the village, the town, the province and the region to the level of the state. Hence the state is spatialized in both 'sitting above' and 'containing' its subjects. In return, the state itself is encompassed by the 'global', often referred to as the 'international community'. In this vertical topography of power, institutions of global governance are perceived as located 'above' states. The 'global' is often referred to as the superordinate scalar level.

The politics of space, we argue, is not the exclusive terrain of established (internationally recognized) state authorities. Political actors in conflict mimic vertical encompassment in many ways. Conflict is about the struggle over the meaning of space. In international relations and conflict studies theory, space has long been understood simply as the site where violence 'takes place': as territory to be fought over. From this perspective, space is primarily seen as a resource to be exploited and owned. Since the 'spatial turn' in the social sciences, however, there has been a growing interest among conflict scholars in studying the connections between space, meaning and power. Taken together, this results in the claim that any analysis of conflict must not focus on discursive representations of legitimacy, authority and superiority alone but include a spatial and practice-oriented approach by analysing the often unmarked, implicit signifying practices through which discourses are made socially effective, for it is believed that it is through these everyday practices that 'bodies are oriented, lives are lived, and subjects are formed' (Ferguson and Gupta 2002: 984). The underlying assumption of a spatial analysis of conflict (and peace, for that matter) is that architectural forms and our built environment (e.g. monuments, bridges, public squares, border crossings, shopping malls and churches) have political and economic impacts on the social systems in which they operate. Buildings, as argued by Montgomery (2008), offer cues suggesting how people should act. They tell us about our relationships with one another. The idea then is to read society through its built environment.

When we try to 'read war' through border landscapes and border architecture we see multiple spatialities at work. Guiding our empirical analysis is the fine-tuned analytical vocabulary of a spatial politics of place, scale and mobility. With the 'politics of place' we refer to the ways in which parties in conflict aim to turn 'spaces' into 'places'. Drawing on cultural geography, we emphasize the conceptual difference between 'space' and 'place'. As explicated by Gieryn (2000: 465), 'Place is space filled up by people, practices, objects and representation.' Space is constantly claimed and framed by actors in their effort to produce a 'sense of place' (Cresswell 2009: 169). Places are imbued with meaning. Actors in conflict seek to strategically manipulate and resignify the meaning of certain spaces through, as we will see, deploying certain imaginaries and practices. The main dichotomies expressed in the Abkhazian-Georgian Conflict are those of separation/access, permanence/temporality and 'war as won' versus an imaginary of the Abkhazian-Georgian case as a 'frozen conflict'. Closely intertwined with a politics of place is the second spatiality: the 'politics of scale'. Central to a politics of scale are the negotiations and struggles between different parties in conflict to reshape or maintain the scalar spatiality of power and authority. Separatist organizations such as the Liberation Tigers of Tamil Eelam, the Kosova Liberation Army and the MNLA engage in scalar strategies such as scale-jumping, turning 'regional' into 'national', to expand their power. The deployment of scale frames is another strategy. Non-state actors, for instance, frame their struggle within a universal/global framework to advance their cause. Conversely, opponents of such claims may emphasize the national interest and notions such as national identity and sovereignty to counter interference from global governance institutions. The third spatial strategy of relevance for this study is the 'politics of mobility'. An essential component within the repertoire of spatial strategies in conflict, mobility refers to 'the material or virtual movability of individuals and objects within and between places, and through space-time' (Leitner, Sheppard and Sziarto 2008: 165). For Sheller and

Law, the 'new mobilities paradigm' seeks to capture how 'All the world seems to be on the move', emphasizing the located and materialized nature of mobility, and (of key importance to our case) associated immobilities (Sheller and Law 2006: 209, cited in Leitner et al. 2008: 165).

In the following sections we examine the signifying practices through which these contentious imaginaries of state have been made socially effective. First, we discuss contentious imaginaries of statehood, their underlying political functionality and their respective propagators. Second, we address the ways in which forms of state spatialization relate to the making and unmaking of identities, citizens and subjects. What stands out in the analysis of the Abkhazian-Georgian Conflict is the impossibility of dissecting the diverse spatialities at work. In line with Leitner et al. (2008) we aim to show the variety, co-implication and alternatability of strategies of place, scale and mobility in violent conflict. In the final section we will discuss how this analysis is significant to the field of conflict studies in the novel ways it addresses expressions of power.

Spatialities of contested statehood in the Abkhazian-Georgian conflict

Since 1993 the Inguri River has acted as the border separating Georgia and Abkhazia. The 213 km-long river originates in the Caucasus Mountains and flows into the Black Sea. It is of key importance to the power supply of Georgia, as the Inguri Dam with its accommodating power station ensures 5.46 billion KWh of electricity a year on average (Democracy and Freedom Watch 2012). However, the best-known part of the river is where the Inguri Bridge connects Georgia with Abkhazia.

During the separation war of 1992–1993, many ethnic Georgians had to flee Abkhazia. Some fled overseas, some went into or over the mountains but most fled over the Inguri River, making the bridge a prop of the Abkhazian-Georgian Conflict. The bridge is the most important legal land crossing point out of the six that currently exist between Abkhazia and Georgia. It has long been the only legal crossing point over the river, but in recent years four additional crossing points for use only by pedestrians have been established in the lower and upper Gali region. The fifth newly opened crossing, at Lekukhona/Alekumkhara, was specifically designated to serve vehicle crossings for the Inguri hydroelectric power station's employees (UN 2014). While the Abkhaz refer to the border as the actual state border between Georgia and Abkhazia, the Georgian authorities refer to it as the administrative border line (ABL), a label that is adopted by the major part of the international community.[2] The border opens around 08.00, depending on the border guards who are present, and closes at 19.00 in the evening. These hours are indicated on a small sheet of paper in Russian on the outside wall of the barracks where passports are checked.

The Abkhazian side of the border is highly militarized and governed as an actual state border. The Georgian side is, although strictly guarded, not treated as a state border but as military checkpoint.[3] If someone wants to enter Abkhazia from Georgia, they first have to pass a Georgian checkpoint. This is manned by Georgian military police who have to control the material movability of individuals and objects across the ABL, ensuring that the route only carries authorized traffic. The way in which the Georgian authorities still consider Abkhazia as a region within its own borders is expressed through the deployment of imaginaries and practices which render the border temporary and informal. The routines of bureaucratic practice that individuals go through when entering Abkhazia from the Georgian side are thus made as informal as possible. Individuals who want to cross the ABL have to show their passport, write down their identification information and legitimize their visit to Abkhazia.

This imagined accessibility is not only performed at the Georgian checkpoint on the Inguri Bridge but also on road signs throughout Georgia. When travelling towards the east of the country, Sukhumi (the capital of Abkhazia) shows up on highway signs as if it is simply a city further down the road. What becomes evident here is how a variety of spatialities is co-implicated in complex ways. The example of the road signs illustrates these complexities, as authorities deploy imaginaries and practices that, while centred on place-making ('Sukhumi is home'), at the same time (re)work mobility ('Sukhumi is accessible') and scale ('Sukhumi is part of the state of Georgia'). By creating an atmosphere in which nothing has changed, and by deploying a spatial politics of 'wholeness' and 'accessibility', these road signs reinforce the discourse of an Abkhazia that is still under Georgian authority. The intentionality of this is exemplified by the fact that the signs have been placed quite recently, certainly after 1993.[4]

Just before the Inguri Bridge there is a last Georgian military bunker with two armed soldiers standing on patrol, observing the Abkhazian side of the border. Next to the road there is a small military garrison with turrets and small bunkers watching over the Inguri River. Once the military garrison has been passed, one has to cross the Inguri Bridge in order to arrive at the Abkhazian side of the river.

A bridge, unbridging

The 870 m-long Inguri Bridge is the main legal border crossing between Georgia and Abkhazia for both vehicles and pedestrians. It was built by German prisoners of war and construction lasted from 1944 to 1948 (Jeska 2004). During construction of the bridge – also known as 'the bridge of the Germans' – 13 German prisoners of war lost their lives. Their bodies are buried in a mass grave near Zugdidi. Today the bridge looks desolate in the almost 1 km stretch of no man's land between Abkhazia and Georgia. Horse carriages transport goods and people from one side to the other. When looking towards the Abkhazian side of the bridge, one can see the Abkhazian flag wave above the hills and trees. When looking back towards the Georgian side, the bunkers with their Georgian flags look more threatening than welcoming. The bridge itself is run down, showing large holes in its asphalt deck.

The United Nations Development Programme (UNDP), with support of EU member states and the United States Agency for International Development (USAID), made plans to renovate the bridge, emphasizing how its deteriorating state could cause dangerous situations in the near future. However, the Abkhazian authorities stated that they wanted to carry out the reconstruction as they claimed that the bridge was on their soil. The Georgian authorities evidently opposed this claim, as this would mean recognizing the Abkhazian authorities. This dispute has resulted in a stalemate. As of late 2014, a number of international organizations had reserved funds for the restoration of the bridge. USAID reserved US$500,000, while the UNDP also focused on renovation as part of the Human Security and Social Integration in Georgia project funded by countries such as Norway, Romania and a number of Georgian municipalities (Romanian MFA 2012, 2; UNDP 2009; USAID 2013: 2). Neither the UNDP, USAID and EU renovation projects nor the Abkhazian plans have materialized so far due to clashing ownership claims regarding the bridge (Interview, Marshania, Tbilisi, 2013). It is hard to miss the symbolic power of this deadlock. As the stand-off continues, the actual material device 'bridging' the opposing factions is slowly but steadily eroding.

The way in which the Georgian authorities have deployed a politics of scale, by allowing 'international' institutions such as the UNDP and the EU to take control of the renovation of the bridge, is also telling. By doing so they emphasize their narrative of the Abkhazian-Georgian Conflict as a 'frozen conflict', requiring the involvement of international intermediaries, and, by implication, as something temporary, unresolved. Conversely, the Abkhazian authorities, scale-jumping from 'province' to 'state', performing an international state border as well as claiming the exclusive responsibility for reconstructing the bridge, enact the narrative of the conflict as having ended (and been won). To add to this complexity, another key player, Russia, also deployed a politics of scale. As we will see below, Moscow's involvement increased from 2008 onwards. Between 1993 and 2008, mainly Russians manned the UN peacekeeping mission that patrolled a buffer zone on the border between the two sides. However, in August 2008, during the war between Russia and Georgia over South Ossetia, Russian troops moved through Abkhazia and pushed into Georgia proper, effectively using the region to open another front with Georgia. Meanwhile, Abkhazian forces drove Georgian troops out of the only area of Abkhazia still under Georgian control – the Kodori Gorge. After the 2008 conflict, Moscow declared that it would formally recognize the independence of both South Ossetia and Abkhazia. In October 2008, Russia pulled its troops back to the Abkhazia–Georgia border but stationed a large force in the breakaway republic, with the agreement of the Abkhazian Government. Scale-jumping from 'international' to 'state', Moscow then also vetoed an extension of the UN peacekeeping mission, and in April 2009 it signed a five-year agreement with Abkhazia to take formal control of its frontiers with Georgia proper (BBC News 2015).

Behind the tinted glass: Russia's quasipresence

When arriving on the Abkhazian side of the bridge, individuals are subjected to a second border-crossing routine. First there is a preliminary passport check before one enters a path with fences and barbed wire on each side. After a short walk of about 100 m along the fenced-off path is the final and official passport control.

This process generally takes a significant amount of time due to the many ethnic Georgians (mainly elderly) who travel between the Gal(i) region in Abkhazia and Zugdidi in Georgia where they receive treatment in hospitals, obtain their medicine and/or visit family. The passport check is performed in a small, improvised 'container booth' with tinted windows, which make it impossible to see who is on the other side. When crossing the border in May 2013, one would have noticed the strong Russian accent of those inside the booth pointing to the possibility that border management was performed by Russian soldiers, an assumption that is further corroborated by the International Crisis Group (ICG). In a 2013 report, the ICG stated that 'During a recent entry by Crisis Group, one Russian and one Abkhazian official manned the booth, with the Russian clearly in charge – though the Russians at the border wear uniforms identical to the Abkhaz, without visible Russian insignia' (ICG 2013: 6).

The darkened windows and lack of Russian 'state paraphernalia' present us with another metaphor: an Abkhazian statehood carefully and quasisecretly monitored by Russian powers. Since the signing of the 2009 border agreement by Moscow and Sukhum, Russia has been the main power-broker involved in the securing of the Abkhazia–Georgia border. Since then, border security management has been subject to a complete makeover as Human Rights Watch (HRW) notes:

> until August 2008 there were four or five official crossing points over the Inguri River and about a dozen unofficial ones, particularly when the river's water level was low. Rules for crossing the administrative border, though arbitrarily applied, were more relaxed.
>
> (HRW 2011: 48)

The 2009 agreement was signed as a temporary measurement until Abkhazia and South Ossetia were able to upgrade their border security themselves (Corso 2009). According to ICG, the agreement was seen by the Abkhaz as a 'symbolic blow', as the former Abkhazian president, Sergei Bagapsh, had been insisting that Abkhazian forces under the formal command of Sukhum would be the 'frontier forces' with the Russians merely assisting (ICG 2013: 5).

Since then, however, Russia's control over Abkhazia's borders has continuously been stepped up. In 2010 the agreement was expanded to maritime borders and Abkhazia's 215 km-long sea border patrolled and controlled by Russian border patrol ships (RIA Novosti 2010). Almost four years after the signing of the border agreement, ICG stated that

> Russia has clearly solidified its security presence in Abkhazia over the past five years, flouting the commitments it made in 2008 to pull back its troops to their pre-war locations, claiming that the agreements are no longer valid because of the 'new realities' created by diplomatic recognition.
>
> (ICG 2013: 5)

Symbolically, in July 2012, the OSCE Parliamentary Assembly passed a resolution in which it now referred to Abkhazia and South Ossetia as 'occupied territories'.

The violent conflict erupting in Eastern Ukraine in early 2014 and the resulting geopolitical stand-off between NATO and Russia have scaled up the tensions between Georgia and Abkhazia to an international level and to what some refer to as the 'New Cold War' (Sadri and Burns 2010: 126). In November 2014, Abkhazia's president, Raul Khadzhimba, and Vladimir Putin signed a military agreement to create a new joint force of Russian and Abkhazian troops. The treaty appeared to be Putin's response to a cooperation deal, named the Association Agreement, signed earlier in the year between Georgia and the EU (EU 2014). Moscow's 'counterdeal' was meant to increase Russia's military control of the Black Sea region, which has long been of strategic importance to Moscow. Putin said that the agreement was to prevent NATO (North Atlantic Treaty Organization) warships from developing bases on the Crimean Peninsula, which is home to Russia's Black Sea fleet. In addition, the Kremlin website stated that the agreement envisaged a joint defence and security space, and stipulated Russian 'protection of the state border of the Republic of Abkhazia with Georgia' (*Guardian* 2014). Not surprisingly, the EU's foreign policy chief, NATO's secretary general and the US State Department condemned the treaty and expressed support for Georgia's sovereignty.

The way in which the upscaling of the Abkhazian-Georgian Conflict from a 'separatist' to an overtly 'geopolitical' conflict is going to find a spatial expression in border security practices remains to be seen. So far, power-brokers such as Russia and NATO have opted not to 'show presence'. At the time of our research (prior to the 2014 Ukraine War), the sensitivity of the Russia–Abkhazia arrangement was expressed in the way border-crossing routines were managed, with Russian soldiers quasicovertly manning the booths and Abkhazian soldiers mainly acting as observers. However, this practice may change as a result of the new spatial politics of scale.

State, space and subjectivity

For most inhabitants of Abkhazia the border is symbolic, as ethnic Abkhazian people rarely cross the Inguri border into Georgia. Of the selection of ethnic Abkhazian people living in Abkhazia that were interviewed for this research, no one claimed to have ever crossed the Inguri border, claiming that the same applied to their family members. All respondents stated that they had never been in proper Georgia and also were not planning to go there in the near future. They also stated that they did not encounter Russian soldiers in their cities. Basically, the military Russian presence and pressure are primarily felt in the border region of Gali, where mainly ethnic Mingrelian Georgians live. This finding is backed up by the ICG:

the thousands of Russian troops tend to keep a low profile in major towns. In the course of a week's visit to Abkhazia and hundreds of kilometers of travel, Crisis Group encountered only a few Russian 'border guards' at the administrative border line (ABL) and a lone military cargo truck. Some locals said this may be a deliberate strategy, probably designed to minimize incidents or creation of an 'occupation atmosphere'.

(ICG 2013: 20)

As the ICG pointed out, the exception is the heavy Russian military and Russian Federal Security Service (FSB) border guard presence along the ABL on the edge of the Gali district. After several years of work along rugged, swampy or otherwise difficult terrain, these forces have 'demarcated' what in Soviet times was merely an unmanned administrative line and sealed off the boundary with concertina wire and trenches.

The spatial discourse expressed by the Inguri border is therefore not immediately associated with, or purely about, physical security. For many Abkhaz it mainly responds to the safeguarding of their Abkhazian identity, as the border is (one of) the spatial dimensions of Abkhazian statehood. As argued by Foucault in *Discipline and Punish* (1977), the essence of the power of the modern state lies in its capacity to distribute bodies and partition space. This is then also one of the most fundamental aspects of state-building: the establishment of a topography of power through bureaucratic practices. The state is performed through the demarcation of territory, staging 'ownership' and, literally, flagging difference, and so reinforcing forms of subjectivity.

The establishment of physical control over territory had actually not been fully accomplished until 2008. Although the Abkhazian authorities claimed to have won the war in 1993, they still lacked control over some parts of what they imagined to be Abkhazian territory. As stated earlier, the Kodori Gorge in northeastern Abkhazia, for instance, was not brought under Abkhazian control until 2008 when it was seized by military means. Another feature of the modern state refers to its governing capacity in the realm of the biopolitical – that is, to control the life and death of the population, and, important in this case, the capacity 'to keep people in their places'. This mechanism of government speaks directly to what we earlier identified as the politics of (im)mobility.

With the arrival of 'Russian presence' from 2009 onwards, border checkpoints have been fortified and 'officialized' on the Abkhazian side (Interview, Marshania, Tbilisi, 2013). As Newman argues, the stronger the barrier function of the border, the more powerful the imagined, the more abstract the narrative of what is perceived as lying on the other side. Perceptions of borders usually focus on what exists on the other 'invisible' side of the line of separation. Borders exist in our mind by virtue of the fear we have of the unknown of the 'there' and which, in turn, causes us to stay on our side of the border in the 'here'.

(2003: 20)

Respondents argued that most ethnic Abkhaz feel that the stronger and more formal the border is, the safer they are (Interview, Marshania, Tbilisi, 2013). In line with Newman's argument, the finding that to the average Abkhaz the border is primarily an identity marker was at times supplemented by narratives of insecurity related to collective memories of paramilitary groups crossing the Inguri River into Abkhazia, trying to destabilize the border region by planting mines during the administration of Shevarnadze and Saakashvilis (Interview, Marshania, Tbilisi, 2013). References were also made to how the border region had to deal with criminal organizations transcending the Abkhazia–Georgia border. Before 2008 it was relatively easy for these organizations to cross the border, intimidate people living in the border region and smuggle goods from one side of the Inguri River to the other (Interview, Marshania, Tbilisi, 2013).

Tight control: Passportization

While for the ethnic Abkhazian people the border seems to be mainly of symbolic importance as it enforces the imaginary of an independent state, for the ethnic (Mingrelian) Georgian people living in Abkhazia, it has a far more direct impact on their everyday lives. Prior to the War in Abkhazia of 1992, people identifying as Georgians made up nearly half of Abkhazia's population. As a result of the war about 250,000 Georgians fled the breakaway republic, reducing its population from about 525,000 before the war to an estimated 243,000 in 2012 (ICG 2006: 19; Autonomous Republic of Abkhazia). Of all Georgian refugees, some 40,000 returned to Abkhazia, mostly to the Gali border district.[5] This group of returnees includes people commuting across the cease-fire line and migrating seasonally in accordance with agricultural cycles, with families often maintaining two residences (IDMC 2014: 51). Many of the ethnic Georgians who had to flee their homes because of the violence are now living in the border region of Samegrelo on the Georgian side of the ABL, near the town of Zugdidi, in artificial villages (Swiss

Cooperation Office South Caucasus 2012). Even since the war there have been short episodes of violence during which people fled Abkhazia to Georgia. According to the Internal Displacement Monitoring Center (IDMC), as of December 2014 there were up to 206,600 IDPs in Georgia. The IDMC report of 2012 states that the living conditions of returnees remained precarious: 'Despite road repairs, infrastructure construction and humanitarian assistance in Abkhazia's Gali district, returnees faced poor housing conditions, insecurity and limited access to basic livelihoods and services' (IDMC 2012). Notably, like most monitoring organizations, the IDMC does not recognize, and hence does not count, Abkhazia or South Ossetia as separate from larger Georgia.

One of the reasons why the Abkhazian authorities enforce tight border control is to prevent the return of Georgian IDPs (Interview, Marshania, Tbilisi, 2013). Many ethnic Georgians tried to cross the Abkhazia–Georgia border by avoiding the checkpoints, but this has become significantly more difficult since the arrival of the Russian border guards:

> The sealing of the ABL has left many locals who do not possess Abkhazian passports feeling increasingly isolated, fearing loss of contact with relatives on the Georgian-controlled side. The new regime has also led to the deaths of several critically ill patients lacking permits to cross into Georgian territory.
>
> (ICG 2013: 19–20)

Ethnic Georgians residing in the Gal region therefore want the border to be as transparent and open as possible (Interview, Marshania, Tbilisi, 2013). Article 6 of the citizenship law of Abkhazia allows dual citizenship only to people of Abkhaz ethnicity, while all others 'have the right to obtain citizenship of the Russian Federation only' as their second citizenship. According to HRW (2011: 36), this provision is clearly intended to deny ethnic Georgians in the Gal district the right to retain their Georgian citizenship when acquiring Abkhazian passports. As the material manifestation of Abkhazian citizenship, the Abkhazian passport serves as the portal to a multitude of rights. Abkhazian citizenship is a mandatory prerequisite in order to acquire the right to work in the public sector, to participate in elections and hold public office, and to enjoy property rights. The Abkhazian passport is furthermore necessary to receive a high-school diploma and to use the simplified procedure to apply for a travel permit to cross the ABL (HRW 2011: 31–37).

The issue of citizenship status not only causes problems within the borders of Abkhazia but is also intertwined with issues concerning the freedom of movement of ethnic Georgians. In May 2013 the Abkhazian authorities suspended the issuing of Abkhazian passports to ethnic Georgian residents of the Gal region, fearing the process might result in – as some Abkhazian officials put it – the 'Georgianization of Abkhazia' (Civil Georgia 2013). In line with this, the Forum of People's Unity of Abkhazia, argued that the issuing of Abkhazian passports to Gali-Georgians was to the 'detriment [of] national security' and was fraught with risk of 'losing sovereignty and territorial integrity' (Civil Georgia 2013: 1). Stanislav Lakoba, the secretary of the State Security Council of Abkhazia, told lawmakers that distributing passports to ethnic Georgians without strict observance of the law would make Abkhazia 'explode' from within (Civil Georgia 2013: 1).

The above examples of the politics of immobility, such as increased border control and the restriction on naturalization, seek to regulate the movement and activities of particular groups in specific places through the creation of laws and regulations, a practice that Anderson coined 'tight control' (2010: 106). Central to a politics of (im)mobility are the struggles between various conflict parties to determine the movability of certain categories of people. Such a politics revolves around the power to enclose, move, contain and circulate. As we will see below, the various parties to the conflict used nationality law, and in particular its materialization into passports, as a biopolitical technology of government.

In the broader context of the post-Soviet space, and Georgia in particular, granting citizenship to large numbers of people, usually an ethnic or linguistic minority, has been used by Russia as a foreign policy tool. In the aftermath of the Abkhazian-Georgian Conflict, Russia issued passports to residents of Abkhazia (as Abkhazian passports cannot be used for international travel). By 2006 more than 80 per cent of the Abkhazian population were in possession of a Russian passport. As Russian citizens living abroad, they do not pay Russian taxes or serve in the army but are entitled to retirement pensions and other monetary benefits (ICG 2006). This 'passportization' evidently caused outrage in Tbilisi. The Georgian Foreign Ministry issued a statement emphasizing that the Abkhaz are citizens of Georgia and calling the passport allocation an 'unprecedented illegal campaign' (Khashig 2002). In the summer of 2011 the Georgian authorities, too, embarked on the use of citizenship rights and passports as a tool to exert population control. The Georgian Parliament adopted a package of legislative amendments providing for the issuance of what became known as 'status-neutral passports' to residents of Abkhazia, which allowed travel abroad and entitled its holder to social benefits in Georgia proper. Allegedly, as an interesting experiment in 'despatializing the state', these documents were stripped of any symbols of Georgia, only showing a neutral registration and individual number. Although no information is available about the actual number of documents issued, 'status-neutral passports' have been recognized by the USA, Japan, Poland, Israel and the Czech Republic, as well as by a number of smaller countries. Not surprisingly, both Abkhazian and Russian authorities were quick to voice their disapproval, calling the passports a 'cunning ploy' by pointing out that they contained imprints of Georgia's country code and the Georgian Interior Ministry as the issuing body (Rinna 2013). The Abkhazian authorities threatened to expel any international organization within its borders that accepted neutral passports. Back in 2008, President Saakashvili deployed yet another variation of immobility politics (isolation) by signing into law legislation that spelled out restrictions on free movement and economic activities to what were considered the 'occupied territories' of Abkhazia and Tskhinvali (South Ossetia). According to the law, foreign citizens should enter the regions through no other country than Georgia proper, making the Inguri Bridge the only official entry point into Abkhazia. It also banned air, sea and railway communications, as well as international transit via the region, mineral exploitation and money transfers. The law was stated to remain in force until 'the full restoration of Georgian jurisdiction' over the breakaway regions was realized (The Law of Georgia on Occupied Territories 2008: 1–8).[6]

In the earlier sections of this chapter we discussed the co-implication of a politics of place, scale and mobility by emphasizing their simultaneous deployment. In the second section we also show evidence for how spatialities can be strategically alternated. While interfering in Abkhazia's border security management, the Russian authorities opted to 'not show presence' (or merely a subtle quasipresence) and hence to not rely on a politics of place, probably to minimize what we might call an 'occupation effect'. Instead, they relied on a politics of scale and mobility, issuing Russian passports to residents of Abkhazia as a tool to exert control over the population. The Georgian authorities, in turn, also embarked on a politics of (im)mobility through isolation and passportization (but in a 'neutral' style, erasing state symbolism). This strategic alternation, however, resulted from Tbilisi's limited spatial repertoire. As we have seen, its place-making capacities went as far as the military checkpoint bordering the rusty Inguri Bridge.

Conclusion: Post-war landscapes

The field of peace and conflict studies has produced a wealth of material about violent conflicts and their legitimation through discursive representations. In this chapter we argue for the need to enrich this analysis by including a spatial and practice-oriented approach. Building on the work of Ferguson and Gupta (2002), we aim to place the signifying practices through which discourses are made socially effective at the centre of our analysis. Underlying this approach is a Foucauldian understanding of power not as something that is 'owned' by certain actors but as 'relational': as a strategic complex relation we are in. The journalist Philip Gourevitch rearticulated this idea succinctly by arguing that 'power consists in the ability to make others inhabit your story of their reality' (1998: 48). In fact, what we argue for in this chapter is to study how certain 'stories' materialize in concrete and tangible entities with spatial properties. We are interested particularly in the story of contested statehood, and an understanding of the state as an imagined entity that exists only by virtue of it being performed (statehood as practice) and spatialized (statehood as materiality). The 'acting out' of the border is key to spatializing the state.

In our analysis of the Abkhazian-Georgian Conflict we have seen how different parties and stakeholders aim to exercise control through the management of space. Through a close reading of border design, and bureaucratic border crossing rituals, routines and regulations, we mapped the operation of co-implicating spatial strategies of place, scale and mobility. This approach allowed us to read this conflict, and the 'rules of the game' upon which it is constituted, through its borderscapes. Geography matters. Borders and landscapes are not 'just there', they are produced and contested, and as such it is through the management of space that actors 'do things' to others (contain, support, welcome, protect, ban, include, keep out, starve). For through spatial politics populations are 'subjected' (at times unknowingly) into certain positions and roles. This is how bodies are sorted, oriented and disciplined, and how subjects are formed.

Post-war landscapes are telling. They tell us stories. Violence leaves traces. And the memory of violence is not just embedded in narratives and testimonies but also inscribed onto space in a variety of settings: road signs, bridges, border-inspection booths, concertina wires, trenches, passports. These landscapes are never stable or fixed, but are contested and renegotiated. It is by studying their (re)design that we can learn about shifting topographies of power and work towards a more comprehensive analytical vocabulary of violent conflict.

Notes

1. Fieldwork was conducted from February to June 2013 in a variety of locations in Georgia and Abhazia. Data-collection strategies included in-depth interviews, participant observation and spatial analysis (see Venhovens 2013).
2. The Republic of Abkhazia is recognized by only a handful of states, including Russia, Nicaragua, Venezuela, Nauru, and also South Ossetia, Transnistria and Nagorno-Karabakh.
3. A (military) checkpoint is a place where (military) police check pedestrian and vehicular traffic in order to enforce control over circulation by imposing control measures and other laws, orders and regulations. It can therefore be an intrastate form of spatial control, while a border checkpoint focuses on controlling the circulation of pedestrian and vehicular traffic between states.
4. The same applies to the capital of South Ossetia, Tskhinvali.
5. According to the Abkhazian authorities, an estimated 60,000 people have returned to Abkhazia (*Abkhaz World* 2009).
6. The Law of Georgia on Occupied Territories, Tbilisi, 23 October 2008.

References

Amoore L (2009) 'Algorithmic War: Everyday Geographies of the War on Terror', *Antipode* 41(1): 49–69.

Abkhaz World (2009) Abkhazian Refugee Situation. Available at: http://www.abkhazworld.com (accessed 2 May 2013).

Anderson J (2010) *Understanding Cultural Geography: Places and Traces.* Abingdon: Routledge.

BBC News (2015) Abhazia Profile: Overview. Available at: http://www.bbc.com/news/world-europe-18175030 (accessed 26 April 2015).

Civil Georgia (2013) Sokhumi Suspends Issuing Abkhaz Passports to Ethnic Georgians. Available at: http://www.civil.ge/eng/article.php?id=26053 (accessed 14 May 2013).

Corso M (2009) Georgia: Russian Border Guards in Abkhazia, South Ossetia Pose New Challenge for Tbilisi. Available at http://www.eurasianet.org (accessed 14 May 2013).

Cresswell T (2009) Place. In: Thrift N and Kitchin R (eds) *International Encyclopaedia of Human Geography.* Oxford: Elsevier, pp. 169–177.

Demmers J (2012) *Theories of Violent Conflict: An Introduction.* London and New York: Routledge.

Democracy and Freedom Watch (2012) Enguri Hydro Power Station Is Still in Georgian Hands. Available at: http://dfwatch.net (accessed 31 May 2013).

Duffield M (2002) 'War as a Network Enterprise', *Cultural Values: The Journal for Cultural Research* 6: 153–156.

Elden, S (2006) 'Contingent Sovereignty, Territorial Integrity and the Sanctity of Borders'. *SAIS Review of International Affairs* 26(1): 11–24.

EU (2014) 'EU and Georgia Adopt Association Agreement', 26 June 2014, EU External Action Press Release 140626/05.

Ferguson J and Gupta A (2002) Spatializing States: Toward an Ethnography of Neoliberal Governmentality. *American Ethnologist* 29(4): 981–1002.

Foucault, M (1977) *Discipline and Punish: The Birth of the Prison*. London: Vintage Books.

Gieryn TF (2000) A Space for Place in Sociology. *Annual Review of Sociology* 26: 463–496.

Gourevitch P (1998) *We Wish to Inform You That Tomorrow We Will Be Killed with Our Families: Stories from Rwanda*. New York: Farrar, Straus and Giroux.

Guardian (2014) Georgia Angered by Russia-Abkhazia Military Agreement. Available at http://www.theguardian.com/world/2014/nov/25/georgia-russia -abkhazia-military-agreement-putin (accessed 23 March 2015).

HRW (Human Rights Watch) (2011) *Living in Limbo: The Rights of Ethnic Georgian Returnees to the Gali District of Abkhazia*. Available at: www.hrw.org/sites/ default/files/reports/georgia0711LR.pdf (accessed 5 July 2015).

IDMC (2012) *Internal Displacement in Europe, the Caucasus and Central Asia*. Available at: http://www.internal-displacement.org/publications/global-overview -2012-europe-central-asia.pdf (accessed 1 July 2013).

IDMC (2014) *Global Overview 2014: People Internally Displaced by Conflict and Violence*. Geneva: IDMC. Available at: http://www.internal-displacement.org/ assets/publications/2014/201405-global-overview-2014-en.pdf (accessed 30 March 2015).

ICG (2006) *Abkhazia Today*. Europe Report. Report No. 176, 15 September. Brussels: ICG.

ICG (2013) *Abkhazia: The Long Road to Reconciliation*. Europe Report. Report No. 224, 10 April. Brussels: ICG.

Interview (2013) Rusiko Marshania, member of the IDP Network 'Synergy' Tbilisi, and Medea Turashvili, analyst at International Crisis Group, Tbilisi, 31 May 2013.

Jeska A (2004) Die Brücke der Deutschen. *Eurasisches Magazin*. 26 March. Available at: http://www.eurasischesmagazin.de (accessed 12 January 2014).

Joseph GM and Nugent D (eds) (1994) *Everyday Forms of State Formation: Revolution and Negotiation of Rule in Modern Mexico*. Durham: Duke University Press.

Khashig I (2002) Abkhaz Rush for Russian Passports. *Institute of War and Peace Reporting*. Available at: https://iwpr.net/global-voices/abkhaz-rush -russian-passports (accessed 5 July 2015).

Leitner H, Sheppard E and Sziarto KM (2008) The Spatialities of Contentious Politics. *Transactions of the Institute of British Geographers* 33(2): 157–172.

Lie J (2002) State Fragmentation: Toward a Theoretical Understanding of the Territorial Power of the State, *Sociological Theory* 20(2): 139–156.

Merry SE (2001) Spatial Governmentality and the New Urban Social Order: Controlling Gender Violence through Law. *American Anthropologist* 103(1): 16–29.

Montgomery C (2008) The Archipelago of Fear: Are Fortifications and Foreign Aid Making Kabul More Dangerous? *The Walrus*. December.

Mountz A (2013) 'Political Geography I: Reconfiguring Geographies of Sovereignty', *Progress in Human Geography* 37(6): 829–841.

Newman D (2003) On Borders and Power: A Theoretical Framework. *Journal of Borderlands Studies* 18(1): 13–25.

Ó Tuathail G (2010) Localizing Geopolitics: Disaggregating Violence and Return in Conflict Regions, *Political Geography* 29(5): 256–65.

RIA Novosti (2010) Russia Builds up Sea Border Patrols around Abkhazia, 12 December. Available at: http://sptnkne.ws/yBk (accessed 5 July 2015).

Rinna T (2013) Georgia's Passport Geopolitics. *New Eastern Europe.* Available at: http://www.neweasterneurope.eu/interviews/936-georgia-s-passport-geopolitics (accessed 5 July 2015).

Romanian Ministry of Foreign Affairs (2012) Disbursement of the MFA's ODA budget in 2012. Available at: http://www.mae.ro/sites/default/files/file/2013/pdf/2013.03.07_Bug_AOD_ENGL.pdf (accessed 5 July 2015).

Sadri HA and Burns NL (2010) The Georgia Crisis: A New Cold War on the Horizon? *Caucasus Review of International Affairs* 4(2): 126–144.

Scott JC (1998) *Seeing Like a State: How Certain Schemes to Improve the Human Condition Have Failed.* New Haven: Yale University Press.

Swiss Cooperation Office South Caucasus (2012) *Vulnerable Households' Accommodation and Community Rehabilitation Project in Samegrelo.* Embassy of Switzerland. Available at: https://www.eda.admin.ch/content/dam/countries/countries-content/georgia/en/resource_en_215765.pdf (accessed 5 July 2015).

UN (2014) Status of Internally Displaced Persons and Refugees from Abkhazia, Georgia, and the Tskhinvali Region/South Ossetia, Georgia. Report of the Secretary-General, 7 May. Available at: http://www.unhcr.org/5385a0779.pdf (accessed 5 July 2015).

UNDP (2009) Human Security and Social Integration Programme (HuSSIP). Available at: http://www.ge.undp.org/content/georgia/en/home/operations/projects/crisis_prevention_and_recovery/hussip.html (accessed 5 July 2015).

USAID (2013) USAID Conflict Mitigation in Georgia. USAID Caucasus Factsheet 2013. http://www.usaid.gov/sites/default/files/documents/1863/conflict%20mitigation%20factsheet.pdf (accessed 5 July 2015).

Venhovens MJH (2013) *Separated Memory: The Influence of Spatial Discourses on the Nation-Building Process of the Republic of Abkhazia.* MA thesis. Utrecht University, the Netherlands.

9

Urban Space as an Agent of Conflict and 'Peace': Marginalized Im/mobilities and the Predicament of Exclusive Inclusion among Palestinians in Tel Aviv

Andreas Hackl

Introduction

This chapter explores the situational emergence of cooperation and conflict among Palestinians who engage with the urban space of Tel Aviv, a city often imagined to be an exclusively 'Jewish Israeli' site. Concerned are both Palestinian citizens of Israel and Palestinians from the West Bank who make use of this city and the opportunities it provides. Inclusion and exclusion coexist here in many ways, and Palestinians in Israel must often balance spatial inclusion into 'Jewish Israeli' space with senses of solidarity and belonging that contradict such immersion. While spheres of Jewish-Arab cooperation emerge in Tel Aviv, the processes of inclusion that facilitate them are often paralleled by the recurring emergence of conflict and tension for the individual Palestinian. Mobility, in the form of commuting and boundary-crossing practices, is as much a tool for overcoming marginalization as it is one of its symptoms.

The mundane struggles of Palestinians in Tel Aviv are approached here through the agency that space itself executes in the situational emergence/dissolution of conflict and 'peace'. This spatial agency is embodied by an assemblage of practices, institutions and values that have come to constitute the city and its boundaries. Expanding Ortner's (2006: 147) use of agency as 'projects', I argue that both urban space and the marginalized people therein try to sustain a culturally and politically constituted project and a certain kind of authenticity. While space mediates Palestinians' capacity to act, the Palestinians mediate the power this space exercises. In making use of the educational and employment opportunities in this city, some also reinscribe their alternative histories and identities onto its surface. Tactics of resistance, cooperation and appropriation are only some of the many dealings with power that emerge and coexist within the same set of social, political and spatial relations in this context. The specific analysis of peace and conflict from a spatial perspective put forward in this chapter integrates Harvey's (2006) concept of 'relational space' with Lefebvre's (1991) 'abstract space'. Relational space exists only through the multiple processes and influences that define it across past, present and future. It is thus best suited to explaining the interplay between historic transformations, political projects, and the plural and shifting subjectivities that relate to such space (Harvey 2006: 273–275). However, as relational space is only fixable through its internal and external relations, the power-related representations of an 'abstract space' such as 'Tel Aviv', and the way it is challenged and appropriated by a marginalized 'urban minority', provide an anchor for such shifting and subjective relations. Space then executes an agency that is both abstract and relational. It is one goal of this chapter to reach a better understanding of how relationships between marginal urban subjects and a dominant urban space simultaneously shape projects of cooperation and conflict.

Much of the contradiction in Palestinians' experience of 'Israeli' space emerged from Israel's foundation as a liberal settler state, a modern colonial polity and procedural democracy established by forcibly removing most of the indigenous majority[1] from within its borders and then extending to those who remained a discrete set of rights and duties that the settler community determined (Robinson 2013: 3). The Israeli political economy is characterized by a high degree of political and geographical power centralization, and it distributes power unequally with the help of territorial 'fracturing' of the main social and ethnic groups (Yiftachel 2001). Moreover, Israel is a polity without definable borders – a highly relational space – particularly because of the ongoing expansion of its territory through settlements in the occupied Palestinian territory (Yiftachel 2006).

Despite the fostering of Jewish-Arab segregation, Jewish and Palestinian urbanites interacted and still interact through a variety of relations and dialectic oppositions (Rabinowitz and Monterescu 2007: 2). The Palestinian citizens of Israel have long moved into Jewish neighbourhoods in mixed or Jewish towns, engaging with the space dominated by the Jewish majority quite routinely. Equally, Palestinian workers from the West Bank have moved to Tel Aviv and into other Jewish towns for decades (Kelly 2006; Portugali 1993).

The spatial engagements of Palestinians with Tel Aviv have remained largely unexplored, not least because the city performs an 'identity' within which Palestinians seemingly do not exist. Such a spatial mirage is connected to the dispersion and decollectivization of Palestinians in Jewish Israeli space, which is itself a symptom of urban marginalization. The Palestinians themselves are stigmatized as 'strangers' there and experience the dissolution of 'place', the loss of a 'culturally familiar and socially filtered locale with which marginalised urban populations identify and in which they feel "at home" and in relative security' (Wacquant 2008: 241). As a result, Palestinians in Tel Aviv rarely form visible collectives, but a collectively invisible mass of individuals and clusters which remain unrecognized as 'Palestinians'. Hence their everyday engagements with Tel Aviv are best viewed as 'collective actions of non-collective actors' (Bayat 2013), including the ingenious ways in which the marginalized make use of the 'strong' (De Certeau 2011: xvii). Young Palestinian citizens of Israel may also seek to take control of their own lives in Tel Aviv as they search for a career or individual privacy. However, their experience of this urban space is often paralleled by tension and exclusion.

It is a common misunderstanding that inclusion, as opposed to exclusion, necessarily needs to be a pleasant experience (Butera and Levine 2009). Palestinians' engagement with the political economy of Tel Aviv creates social and economic exchange through exclusive inclusion within which the same set of relations becomes a source of individual empowerment and of deepened marginalization. The Palestinian engagement with Tel Aviv remains largely temporary for a number of reasons, among them estrangement and exclusion, but also their own communal requirements to return to their home areas. One can look at the temporariness of the Palestinian experience in Tel Aviv as a failed 'arrival city' (Saunders 2011) – clusters of Palestinians in the city's cultural, political and economic margins, without the chance (or desire) to establish a sustainable and permanent 'home' in the city. Instead, mobility and commuting create a fluctuating urban presence.

In trying to harvest the opportunities in Tel Aviv, the Palestinians' ability to be socially and spatially mobile is mediated by their marginalized position of departure. This idea is expressed by the concept of 'motility', which integrates spatial and social mobility by looking at the ways entities access and appropriate the capacity for sociospatial mobility according to their circumstances (Kaufmann et al. 2004: 750). The capacity to engage successfully with the urban space of Tel Aviv is then deeply relational, and such engagement produces the shifting subjectivities that relate to such space (Harvey 2006: 273–275). Hence Palestinian mobility into Tel Aviv is a response to systematic exclusion, particularly in the case of Palestinian workers from the occupied territories, but also among citizens of Israel who have suffered from dispossession and socioeconomic gaps, alongside many other forms of inequality.[2] As power mobilizes labour across space, these disadvantages mediate the possibilities of urban inclusion and 'arrival', and individual adjustments become necessary. Indeed, the same relations that fulfil individual needs also entrench 'unfreedom' through a particular assemblage of spatial practices that simultaneously compel and confine (social and spatial) movement (Kothari 2013). Movement hardly equals freedom there but suggests that mobility, inequality and power interact in multiple ways (Glick-Schiller and Salazar 2013; Salazar and Smart 2011; Shamir 2005). And there seems to be an obligation to be circulating, as in many other systems which have recently evolved (Urry 2007: 13). Spatial and social mobility operate within a system of ethnoracial confinement there (Wacquant 2011).

In exchange with these conceptual insights, this chapter will explore the emergence of spaces of conflict, 'cooperation' and co-labouring among Palestinians in Tel Aviv based on four cases: Palestinian workers from the West Bank; Palestinian citizens of Israel in the high-tech sector; Palestinian women from Jaffa who seek employment in neighbouring Tel Aviv; and Palestinians in Tel Aviv who stage political protests and individual actions that challenge marginalization.

Confined workspaces and circular mobility

In order to harvest the opportunities within the state that occupies their land, tens of thousands of Palestinian workers from the West Bank make their way into Israel and into Jewish settlements every day.[3] All that is seemingly for one main reason: to 'bring bread to the family', as they say. The circular movements involved are meant to overcome the impact of occupation and exclusion, as much as they are the very product of the contradictions of the Israeli state and the logics of exclusion that underlie its rule over territory and people (Kelly 2006: 111). Also, the requirement to be apolitical forms part of the economy's exclusive inclusion.

The analysis put forward in this chapter is based on two years of ethnographic research between 2012 and 2014 among Palestinians in Israel, with a focus on Tel Aviv. Such anthropological urban ethnography is predominantly qualitative and builds on richly contextualized accounts of human thought and action; it can focus on people's experiences of social situations in cities to assemble complex social life from them, but urban ethnography should also be the research of the city itself, rather than merely using it as a self-explanatory backdrop of social phenomena (Hannerz 1980: 8–10). Methodologically, this chapter acknowledges the analytical centrality of everyday practices as individual ways of 'operating', which become a locus in which a plurality of relational determinations interact in relation to space (De Certeau 2011: xi). ' "Space" is composed of a polyvalent unity of conflictual programs or contractual proximities' (De Certeau 2011: 117). Jewish-Arab co-labouring at workplaces in Tel Aviv forms such contractual proximity despite underlying conflict.

One hot summer day in Tel Aviv I met a group of Palestinian construction workers for lunch in a shaded space below a building, and somehow the conversation turned to earlier research I had done on Palestinian civil resistance in the West Bank village of Bil'in. 'Look,' one of them said, 'those who protest in Bil'in get tear-gassed and jailed. Those who don't go to Bil'in can work here and make money. What would you do?' He added: 'This is how we have become: we think about earning our living and are left with no energy to do other stuff.'

These workers knew well that political activity could easily lead to a cancellation of their working permit. The conditions of inclusion prefigured their engagement with the 'Jewish Israeli' space of Tel Aviv as 'work'. The regulated workspaces are seemingly detached from the inequality that triggers their initial formation because they remain isolated from the flow of social life and leisure in the city.

At the construction site of a high-rise building in Tel Aviv, I met another cohort of Palestinian workers. After walking across a makeshift stairway into the staff offices in the basement, I was introduced to Fahed, a 47-year-old Palestinian man from the West Bank city of Hebron. His usual work day started at 03.00 in Hebron, from where he took a shared taxi to the nearby Israeli military checkpoint, waited for at least half an hour to pass the security checks, and boarded a bus on the other side towards Tel Aviv. Each night he had to return because his permit did not allow him to stay in Israel overnight. 'Why are you doing it?' I asked. He responded:

Of course it's hard, the waiting at the checkpoint, the journey... But if you ask why, I say it's the salary. In Hebron I would earn less than half... but movement also makes me tired, causes problems in my family. When I come home my child asks for help with homework but I am simply too tired. I miss out on weddings and social events. On Friday, when I come back home, I sleep all day.

The circular movements caused fatigue and disconnected him from social life at home, while his permit also restricted any non-work engagement with Tel Aviv. In another office container, I met the 30-year-old Abdallah from Hebron. As we started to talk, he suddenly took off his helmet and slammed it upside-down on the table. 'You see this helmet? This is our construction site. In here you have a mix of everything, everything goes well. But outside you have politics and you have troubles. Here, what we care about is work.' He said that bringing money to his family took priority, but to achieve that 'you have to shut up and move on'.

In the early afternoon we had lunch with other Palestinian workers and Jewish engineers, some of whom spoke a bit of Arabic. There was a calm and friendly atmosphere around that plastic table as we ate bread and hummus. 'There are no problems here,' said one of the Jewish Israeli employees. 'Politics is outside. If you have bread, there is peace.'

These workspaces are metaphors for 'coexistence' as Palestinian workers remain bound to a highly regulated sphere of inclusion in a political economy that is characterized by spatial diffusion; it depends on Arab 'labour nomadism' and the labourers' ability to follow the changing demand for labour at the bottom of the earnings ladder (Portugali 1993: 14, 72). This dependency emerges from the power asymmetries in the Israeli-Palestinian relationship and is sustained by Israel's unchallenged domination of Palestinian life (Roy 2007: 79). The discussed workspaces may momentarily freeze a relationship that is otherwise defined by power asymmetry and conflict into regulated spaces of 'pragmatic peace'. However, these are places of Israeli-Palestinian co-labouring where cooperation is both limited and facilitated by the underlying structures of inequality and domination.

Between inclusion and outsourcing

In the case of the Palestinian citizens of Israel, power asymmetries may at first sight be less visible than in the occupied territories, but they are no less significant. As Israeli citizens, they may enjoy certain privileges, but citizenship itself is a tool for regulating inclusion and exclusion in Israel. It is systematically stratified along ethnic and religious lines, and legally is highly differentiated (Yiftachel 2002). The Palestinian citizens of Israel face a matrix of walls and glass ceilings that prevent integration or equality in the labour market and elsewhere (Pappé 2011: 245).

The story of the Palestinian citizens of Israel – about 1.4 million, or 17 per cent of Israel's population of 8.3 Million by 2014[4] – began with 150–160,000 Palestinians who had managed to stay in, or return to, what had just become Israel after the Arab-Israeli War of 1948 (Pappé 2011; Robinson 2013: 1). Since then the Palestinian citizens of Israel have also taken part in an 'unending contest over space', facing continued expropriation of land and insufficient urban planning in Arab towns (Pappé 2011: 254; Wesley 2013: 192). With the disappearance of Palestinian urban life after 1948, the culturally familiar arenas for productive exchange were severely limited. One may speak of the 'absence' of a Palestinian Arab city in Israel, as larger Arab towns are better described as overgrown villages, and so-called 'mixed' towns have an overwhelmingly Jewish Israeli character (Rabinowitz and Abu-Baker 2005: 116). In the city of Tel Aviv, such dominant 'character' lends an element of cultural and political exclusion to the agency this urban space exercises.

One of the employment sectors where the glass ceilings for Palestinian citizens are particularly thick is 'high-tech'.[5] There, initiatives aimed to challenge inequality through the proactive inclusion of Arab engineers in Tel Aviv and elsewhere. In 2013 about 1,200 Palestinian citizens of Israel worked in high-tech companies, compared with only 350 in 2008, according to the Nazareth-based organization Tsofen (Maltz 2014), one of those initiatives.

However, such inclusion of Palestinian citizens into a highly competitive sector dominated by Jewish Israeli employees, entrepreneurs and investors faces multiple obstacles. The pursuit of overcoming these followed two main trajectories: training Arab engineers for employment in the centre of the country, and creating work and development opportunities within Arab population areas through 'outsourcing'. Such outsourcing is usually associated with sending jobs overseas and seeks to employ people who are otherwise excluded from the labour market (Gately 2014: 290).

One of the driving forces behind outsourcing in Israel was Inas Said, who founded the Arab-led software company Galil in 2008. This move was triggered by his own experience in Israeli companies. 'I realised that

I was the only Arab in a company that had 3,000 employees and more than 500 engineers.' Then he came up with the idea of outsourcing, forming an 'incubator' in which a 'real high-tech environment' could be simulated. It was a way of shaping a mass of engineers in a safe environment, rather than providing risky opportunities for just a few. Outsourcing into Nazareth's industrial park should create spatial proximity and cultural familiarity between the Palestinian population in Israel and the high-tech-sector, according to Said. It also countered the spatial inequality of resource distribution in the Israeli political economy: only 2.4 per cent of all industrial zones in Israel are located in Arab communities (Orpaz 2015).

However, owing to the lower salaries involved in outsourcing and the desire to go beyond the limitations of the 'incubator', Palestinian-Israeli engineers continue to search for jobs in greater Tel Aviv, the heart of the Israeli 'start-up' economy. In doing so they benefit from the training programmes of the non-profit organization Tsofen, which promotes the integration of Palestinian citizens of Israel into the high-tech industry.

Smadar Nehab, one of Tsofen's founders, had just come back to Israel after a career in Silicon Valley when her company's board suggested that she should search for new talent in India. 'I said to myself: "Why India?" They are in the Galilee (region in Israel). I was aware of the shortage of good talents in high-tech, and I was aware of the absence of Arabs from the workforce.' However, aligning the interests of the Israeli economy with that of Arab-Palestinian engineers was challenging in many ways. 'The industry likes to be homogenous. There are stereotypes, [there is] hatred.' said Nehab.

Moreover, the Palestinian citizens of Israel often lacked the networks and knowledge of Jewish Israeli engineers, and some of these networks were formed during Jewish citizens' mandatory service in the Israeli army, from which Palestinians are exempt. Nehab considers placing Palestinian citizens of Israel into high-tech companies in Tel Aviv to be 'almost impossible', which is why Tsofen built up a high-tech centre in Nazareth too. Besides anti-Arab prejudice and racism, others cite indirect barriers, such as companies' rules not to hire employees who live far away. 'It seems very rational, but Arabs mostly live further away. So they can't work there, unless they live around Tel Aviv... This geographical separation became such a barrier,' said Nehab. Why, then, don't Arab engineers live in Tel Aviv instead of commuting or working in outsourcing?

One of Tsofen's former participants is Faris, a 35-year-old Palestinian citizen of Israel born in the Arab town of Sakhnin. He preferred Tel Aviv over outsourcing 'at home' because of the better salaries, and because outsourcing tied people to the 'incubator' rather than mobilizing them. But workplace inclusion in Tel Aviv also has its price. It begins with the commute – in the case of Faris, four hours a day on public transport, on top of ten working hours. But the real issue lies in his experience of Israeli space and transport.

> When I approach the guard in front of the train station while talking Arabic on the phone they always check me and ask me a few questions. If I don't [speak Arabic], they don't. And recently on the train my phone rang and I picked up, speaking Arabic. They all stared at me and I felt how they went one step away.

The experience of Israeli public space can be unsettling for Palestinians, a tension that appears during the frequent transitions across spatial boundaries. What the individual Palestinian 'shows', like speaking Arabic, and what remains hidden, is part of a complex everyday politics of negotiating 'backstage' and 'frontstage' in these processes (Goffman 1959).

'It is a catch-22,' said Inas Said about Palestinian citizens of Israel who blend into the city of Tel Aviv. 'Their only chance to find a job is to be individualistic. If they start organising, the Israeli state will restrict them with new laws. As an Arab activist, of course I want them to have more rights. But on the other hand, I don't want to be the person causing them to lose the last opportunity they have.' In search for a job or a little rental space in the city, the Palestinian citizen of Israel is often expected to be a 'good Arab', which essentially refers to a sort of non-Palestinian, a persona that is invisible in terms of their Palestinian identity.

Invisibility as a Palestinian Arab helps to avoid tension in the Israeli system of exclusive inclusion, as the female Arab employee and student Dania explained in Tel Aviv:

> At university, the guards always think I am not Arab. But recently, they began to request student cards. They always said hello and were friendly. But when they see my card and my family name, they realise I am Arab and ask me to open my bag. Before, it was as if I was VIP, no one expected that I am Arab.

If not being visibly Arab leads to 'VIP' treatment, eases transitions and avoids confrontations, the same political economy made use of her 'Arab-ness': one of Dania's student jobs was in one of several Tel

Aviv-based companies that employed Arabic-speakers for their regional service desks and sales departments. 'I was there for Saudi Arabia, Kuwait, and so on,' she said. Evading politicization through invisibility as a Palestinian may be one of several tactics by which stigmatized people cope with 'being a problem' (Bayoumi 2008). And this imposed 'problem' intensifies during times of war and polarization, as during the 2014 flare-up in violence between Israel and Hamas in the Gaza Strip.

Recurring political tension destabilizes the fragile balance Palestinians in Israel try to keep. 'I reached a point where I prefer to talk English instead of Arabic. I feel I am in danger,' said Towibah, a Palestinian-Israeli woman who worked and studied in Tel Aviv during the Gaza escalation. And attacks on Arab protestors or agitated verbal abuse were very common during this war, as polarization gradually intensified.

Following the 2014 Israel-Gaza Conflict, a report in the Israeli newspaper *Haaretz* (Maltz 2014) discussed the impact on Arab workers in Jewish companies. 'Dozens of complaints have been received in recent weeks from Arab workers who have either been fired or threatened with dismissal for expressing what she [the attorney] describes as 'thoughts outside the consensus' on social media sites,' the article stated, citing an attorney from the Nazareth-based Worker's Hotline. In the same piece, Smadar Nehab is cited as saying that high-tech companies showed less willingness to hire skilled Arab engineers, stating: 'This is not the time. Let's wait until things calm down.'

However, things never entirely 'calm down' in the lives of Palestinians in Israel, and there are many reasons for not settling down in Tel Aviv, including the absence of appropriate schooling, the marginalization of the Arabic language, and the wider feeling of cultural and political estrangement, as well as economic reasons. Outsourcing is seen as one possible alternative to the exclusive character of inclusion into Israeli workspaces, but it also entrenches the spatialized difference between Jewish and Arab citizens, despite the cited merits. However, if the creation of sustainable urban coexistence and spatial familiarity for Palestinians in Tel Aviv remains difficult, evading the underlying layers of discrimination and inequality through outsourcing seems like a viable alternative to confronting them on a daily basis.

Solid boundaries and imageries of immobility

Another initiative that seeks to ease Palestinians' access to employment opportunities in Jewish Israeli space is the NGO Arus al-Bahr (The Bride of the Sea). It supports local Palestinian women from the Arab neighbourhood of Jaffa who seek appropriate work, a resource to be found in Tel Aviv and other nearby suburbs. Since Tel Aviv's founding in 1909 when it started as Jaffa's 'Jewish garden suburb', it has overshadowed Jaffa economically and demographically in the 1930s, conquered it in 1948 and 'annexed' it in 1950, 'rendering it a dilapidated South Side and perpetuating its political and cultural otherness' until today (Rabinowitz and Monterescu 2007: 12). Although Tel Aviv and Jaffa have since 1949 been one city administratively, they are very different and yet interconnected places.

South of Jaffa's gentrified old town, the office of Arus al-Bahr is on the top floor of an abandoned-looking building. There, I met Safa Younis, the NGO's executive director and founder, who was born in Jaffa in 1975. She was a pioneer in Jaffa because she attended a Jewish school in Tel Aviv at the age of 16. Although this was an unusual decision, it eased her way towards successful inclusion into university life. 'I started to live the life of a minority in Tel Aviv,' she said.

However, for many of her clients the idea of working in Tel Aviv seems like migrating to another country, said Younis. It is clear that both the meanings attached to mobility and the capacities to be mobile are mediated by gender-specific limitations, and that any attempt to overcome such exclusion needs to take these specificities into account. The women whom Younis deals with are often married, have several children and come from a lowly socioeconomic background, many of them having never been formally employed. 'For most, the first reaction is: It's not for me, I am scared,' she said. 'The image is often that they [the Jews] won't accept us, that they will stare at us.' Most of her clients wear a head-scarf and know that they may be discriminated against on the Israeli job market for that reason. Educational gaps, family problems and financial troubles add further insecurity. 'I try to give them dreams, help them to see their own potential,' Younis explained.

Arus al-Bahr prepares women for job interviews and builds links between them and employers. The barriers to overcome include invisible but powerful borders and fear that mark the space between Tel Aviv and Jaffa, according to Younis:

> A lot of women are scared to be attacked in Tel Aviv. One woman told me recently that the furthest place she would work in is the clock-tower in Jaffa [bordering Tel Aviv]. I asked her why, and she said that she heard about one woman being attacked there...There are also fears about being looked at, that they are not good enough in Hebrew, that they are wearing the veil and that it would be strange.

Although administratively one city, the space of Tel Aviv-Jaffa is differentiated by invisible boundaries that mark some Palestinian women's safety zone against the urban space of the Jewish Israeli other. Younis said that they 'always feel like a minority', like 'guests' in Tel Aviv, 'even if we are actually from here'. A the same time, however, 'If a woman here wants to advance, she has to go to these Jewish places in Tel Aviv and so on. She simply has to deal with it.'

One of Arus al-Bahr's earliest clients was 39-year-old Zahie, who ended up working in the client intake section of an insurance company in central Tel Aviv seven years ago. 'Because I was wearing this [pointing at her headscarf], the only jobs I could find in Tel Aviv was work on the phone. No one else would have accepted me with a hijab in Tel Aviv,' she said.

Earlier she had been working as a waitress in the Dizengoff Centre shopping mall in Tel Aviv, not yet wearing a headscarf. Besides her name tag, nothing at that time would have marked her look as particularly Palestinian Arab. Her memories of one work day remain vivid. 'I was working as usual when the manager suddenly came over and took off my name tag. When I went outside Dizengoff Centre I understood why. There was a bombing and groups of people started to shout "Death to Arabs, death to Arabs." ' It was 4 March 1996, when a suicide bomber blew himself up outside the mall, killing 13 Israelis. The fragile balance of co-labouring was upturned and the 'invisibility' of her Palestinianness became the only guarantor of safety – an impossible invisibility for women wearing the headscarf. Conflict and danger can arise unexpectedly, and such unpredictability creates a feeling of vulnerability among Palestinian individuals in Israeli space.

Much later, Zahie joined Arus al-Bahr as a project manager. Together we met one of her clients, the 37-year-old Heba,[6] who was born in the Gaza Strip but lived with her husband and six children in Jaffa under the terms of a temporary residence permit which did not allow her to work. She had been in Jaffa for 12 years at the time we met and her Hebrew was still weak. She joined Arus al-Bahr to find a better job and to gain confidence. 'I need to learn how to express myself better, how to survive in Tel Aviv,' she said. Whenever she has a new job offer in Tel Aviv, she needs someone to accompany her because she 'won't find the street' and is 'scared to get lost'. The many aspects of her marginalization relative to the spaces of the Israeli political economy severely limit her capacity for sociospatial mobility, and consequently her chances for workplace inclusion. These limitations interact with culture and gender-specific ideas about work and public space.

Women like Heba, Zahie and Safa become living imageries of possibilities, and yet it is evident that not all obstacles can be overcome. While sitting in a coffee shop, Zahie pointed towards Jaffa's clocktower, saying: 'For them, this clocktower is the border. Everything beyond is Tel Aviv', and this 'beyond' is one of many gendered insecurities connected to mobility. Initiatives such as Arus al-Bahr then become mediators between multiple marginalized Palestinians and the Israeli political economy, between subjective desires and relational possibilities – a cooperative sphere for negotiating the boundaries between spaces of familiarity and estrangement where 'peace' and tension often flow into each other.

Challenging spatial domination

The discussed processes of workspace inclusion demand a difficult balancing act from Palestinians. As they invest much to overcome marginalization and exclusion, whether through training, commuting or outsourcing, they also need to balance their pragmatic interests with the recurring nature of tension and conflict in Israel's political economy, which often demands depoliticization and 'invisibility' within a sphere of exclusive inclusion.

However, Palestinian individuals and initiatives also challenge domination and exclusion in Tel Aviv, which has ever since its creation been viewed as in opposition to Palestine and Jaffa (LeVine 2005: 6). As Tel Aviv performs its own agency and identity, Palestinians also seek to reclaim their own narratives about this space and its history. Hence they challenge this 'abstract space', which has become a tool of domination as it erases the historical conditions that gave rise to it, hides its own internal differences and imposes abstract homogeneity instead (Lefebvre 1991: 370). Borrowing from Lefebvre's metaphor of 'rhythm' can help one to understand similarity and difference in relational urban space. One can look at Tel Aviv as a space that exercises power as a political and economic centre by homogenizing the polyrhythm of internal diversity. Yet when relations of power (or domination) continuously overcome relations of alliance (or cooperation), compromises can implode, and contradiction and crisis may surface (Lefebvre and Régulier 2013: 105). What is usually kept backstage for the sake of pragmatic opportunities may then surface when Palestinians in Tel Aviv actively bring their alternative narratives frontstage, thereby breaking with the logic of the depoliticization of workplace inclusion.

One alternative rhythm of self is 'performed' by the Palestinian-Israeli artist Anisa Ashkar, who lives in Tel Aviv. When she wakes up in the morning, she contemplates for a while until something meaningful comes to mind, and then she writes onto her face in Arabic calligraphy. 'I started doing it 14 years ago because I am Arab. They always thought I am French, or from somewhere else. In Tel Aviv, people don't get that I am Arab. "Are you Brazilian?" they ask. I wanted to underline that I am Arab. I wanted to have ownership over my own body.'

Ashkar sees Tel Aviv with different eyes and makes sure the city sees her with different eyes too. She inscribes senses of identity onto the surface of a space which otherwise silences them. She does not attend organized protests, saying that 'to survive and breathe here is political. To be an Arab artist in Israel is political. You talk about protests? My face is a demonstration.' The need arose because of external, rather than internal, reasons, Ashkar explained. The problem was not her own self-view but the way Jewish Israelis ignored, suppressed or avoided the fact that she was a Palestinian Arab. 'But it is also to remind myself that I am Arab, because I am constantly in Jewish or foreign space.'

Tel Aviv has also been the site of occasional Palestinian protests, one of which was held in 2014 by the National Democratic Assembly (Balad), an Arab political party in Israel. The demonstration in front of Israel's Ministry of Defence was directed against attempts by the state to draft Christian Arabs in Israel into the army. The sight of Palestinian flags and chanting in Arabic interfered with the dominant rhythm of everyday Tel Aviv, leaving some in shock, some wondering and others angry. 'Die, die Arabs!' shouted one passer-by at the demonstrators. One man spat at them from his car window, while other verbal attacks included 'Go back to Gaza!' and 'Be happy that you are not in Syria!'

Tel Aviv is often seen as a liberal and open city with respect for diversity, but the reactions of passers-by against this protest exemplified how urban political culture fends off collective representations of Palestinian identity. As another agitated Israeli tried to pursue the police to halt the protest, Awad Abdel Fatah, the General Secretary of Balad, said: 'He wonders how the police can allow us to be here. When they see a Palestinian flag in the heart of Tel Aviv they are shocked ... They don't want to see anything except of themselves.' Indeed, 'when "rhythms of the other" make "rhythms of the self" impossible, crisis breaks out' (Lefebvre and Régulier 2013: 105). The recurring character of such 'crises' makes the emergence of sustainable spaces of Jewish-Arab peace and cooperation in Israel increasingly difficult.

Regular political actions also took place at Tel Aviv University (TAU), where about 2,040 Palestinian citizens of Israel studied in 2014. There, student activists organized protests against Israel's policies or military campaigns in the Gaza Strip. Fierce reactions and verbal abuse has also become a routine response here. Another regular event at TAU was a commemoration ceremony of the Palestinian catastrophe of 1948, the *Nakba*. The mere approval of the ceremony unleashed a storm of debate in Parliament. Police forces surrounded the crowd with a fence, a measure of confinement that simultaneously provided protection from the agitated crowds of Israeli nationalists who waved Israeli flags and denounced the *Nakba* as a lie. In fear of being attacked, Palestinian participants were later requested to disperse into the anonymity of the gated campus area before walking off anywhere else individually, 'to be secure'. Invisibility was once again necessary.

Moreover, the university itself stands on land that used to belong to the destroyed Palestinian village of Sheikh Munis, a Palestinian guide told dozens of participants during a *Nakba* tour in Tel Aviv. The tour traced Palestinian history throughout Tel Aviv as the group of students moved between the locations of former villages in the city, among them Manshiyeh, between Jaffa and Tel Aviv. After some background information and a display of photographs, a workshop at the beachfront let the Palestinian students imagine the implementation of a future return of Palestinian refugees to their respective places of origin.

The guide then pointed out the remains of another pre-1948 building on Tel Aviv's seaside boulevard. In crude manner, it has been transformed into a memorial to the pre-state Zionist paramilitary group Etzel. The irony was not lost on the participating students, who seemed furious about the following words written prominently onto the walls of the building: 'Etzel House: In memory of the liberators of Jaffa.' Needless to say, to them, 'liberation' meant occupation. The dominant narratives of history in this urban space had once again fooled them.

Conclusion

Engaging with Israeli urban space poses a difficult dilemma for Palestinians and Palestinian citizens of Israel: it is here where education and employment opportunities are concentrated, which is why individuals make use of them alongside Jewish colleagues at the workplace, or at the university; and these processes of inclusion and mobility may be one way of overcoming inequality and its inscriptions in space. However, they are also a symptom of the systematic exclusion Palestinians face in the Israeli political economy, both collectively and as individuals. Moreover, as we have seen, tension and conflict frequently surface in the everyday engagements of Palestinians with Tel Aviv, thereby limiting the possibility of sustaining spaces of 'peace' and cooperation.

Palestinians aim to overcome inequality by commuting across a whole infrastructure of separation and marginalization, as in the case of Palestinian workers from the Israeli-occupied West Bank. On the way out of the margin, powerful imageries of immobility can solidify the boundaries between spaces of familiarity and those of sociocultural estrangement or political marginalization. This became evident among work-seeking women from Jaffa. Outsourcing opportunities into the Arab periphery of the political economy may seem to be a possible way out of the dilemma, but we have seen that individual Palestinians in Israel continue to pursue employment in Tel Aviv for various reasons, including economic merit and the desire to enjoy an urban lifestyle despite the underlying tension.

The Palestinian 'population' of Tel Aviv fluctuates and maintains a highly temporary presence: the coming and going between Tel Aviv and Palestinian citizens' home towns in northern Israel, the five-hour commute from Israeli-occupied territory in the West Bank, or the difficulties of relating to this space with cultural and political affinity. The spaces of transport used in commuting are themselves agents of peace and conflict, as are the many transitions. This is particularly true during times of war and tension, when a routine of being profiled by security guards and the experience of anti-Arab discrimination only increase the feeling of estrangement in Israeli public space.

Moments of crisis accelerate and intensify the rhythm by which tension and conflict recur and are experienced. When the overall atmosphere heats up, frozen conflicts thaw quickly, only to freeze again when the necessities of everyday pragmatism demand it. But these pragmatic adaptations have their price when the daily routine in Jewish Israeli space becomes difficult to reconcile with individual senses of identity and solidarity. In spaces of Jewish-Arab co-labouring, the Palestinians often face pressures to keep their political beliefs and national solidarities 'backstage'. They may even be 'depoliticized' as an individual non-persona, a 'good Arab', as exemplified by the maxim 'Where there is bread there is peace.'

Most importantly, this chapter explores the agency that space can exercise in the everyday emergence, recurrence and dissolution of peace or conflict. This agency draws on the power differentials that operate in space and are produced by it. Tel Aviv, as I have shown, also performs certain narratives of history and identity, which are challenged by spatial practices. Other examples of urban agency include the branding of places, such as former Palestinian houses, neighbourhoods or destroyed villages; or the way passers-by react to and attack political protests as a form of sanctioning unwanted boundary transgression; and the way Jewish Israeli majority space operates socially during the various national holidays and memorial days, some of which are accompanied by the roar of a siren that transcends every corner of urban space. As most Jewish citizens hold a minute of silence, most Palestinians avoid being in public space during these moments. Such exclusive spatial agency increases the feeling of estrangement. For some Palestinians who merely use Tel Aviv and its opportunities, urban space then becomes a function rather than a place of belonging. Palestinians' own conceptions of Tel Aviv, and the non-fulfilment of their cultural and communal needs in it only further restricts the ability of these citizens of Israel to immerse or 'arrive', while Palestinian workers from the West Bank are forbidden any sort of immersion by the rule of the regulated workspaces that their entry permits dictate.

Yet others challenge the proscriptions of spatial domination by reinscribing their alternative versions of spatial identity and history into this space and onto its surface, whether by individual everyday practices or through coordinated activities. Such highly visible encroachments exist parallel to everyday workplace pragmatism and the daily routines of Palestinian Arabs in Tel Aviv, which certainly also involve peaceful Jewish-Arab interactions and mutual respect. However, as long as inequality and marginalization beneath the surface of such spatial engagements are not addressed, and as long as the history and identity of Palestinians in such spaces remain suppressed by a homogenizing tendency, the balance of everyday coexistence remains a highly fragile one.

Notes

1. Palestinian Arabs comprised 90 per cent of the country's inhabitants in 1922 (Robinson 2013: 2).
2. Such inequality includes some 30 main laws that discriminate, directly or indirectly, against Palestinian citizens of Israel; income gaps and poverty, unequal resource distribution, discrimination in the employment sector and in terms of access to employment, health and land; as well as marginalization of the Arabic language and restrictions on political participation (Adalah 2011).
3. Israel regulates permits based on quota and security screenings of individual Palestinians. In March 2014 the quota of legal working permits was 47,350,

excluding those crossing into Israel without permits (B'Tselem 2014). According to the workers rights NGO Kav LaOved, approximately 30,000 Palestinian workers are currently employed in Israel, most in the field of construction, as well as many in agriculture, industry and service jobs. An additional 30,000 are employed in the Israeli settlements (Kav LaOved 2012).

4. In its population count, Israel does not distinguish between Palestinians in Israel and those in East Jerusalem, over which Israel claims sovereignty, an act not recognised by the international community because East Jerusalem is occupied territory under international law. The 1.4 Million figure excludes most of the 300.000 Palestinians in Jerusalem who are not citizens.

5. High-tech goods and services account for 12.5 per cent of Israel's gross domestic product (GDP) and for half of its industrial exports. Israel leads the OECD in research and development, spending 4.3 per cent of its GDP on it – nearly twice the OECD average (Reuters 2015).

6. Not a real name.

References

Adalah (2011) *Inequality Report: The Palestinian Arab Minority in Israel.* Haifa, Israel: The Legal Center for Arab Minority Rights in Israel (Adalah). Available at: http://www.adalah.org/uploads/oldfiles/features/misc/Inequality_Report.pdf (accessed 5 July 2015).

B'Tselem (2014) *Intl. Workers' Day: No Cause for Celebration for Palestinians Working in Israel.* 30 April. Available at: http://www.btselem.org/workers/20140430 _international_workers_day_2014 (accessed 6 July 2015).

Bayat A (2013) *Life as Politics: How Ordinary People Change the Middle East.* 2nd edn. Stanford: Stanford University Press.

Bayoumi M (2008) *How Does It Feel to Be a Problem?: Being Young and Arab in America.* New York: Penguin Press.

Butera F and Levine JM (eds) (2009) *Coping with Minority Status: Responses to Exclusion and Inclusion.* Cambridge: Cambridge University Press.

De Certeau M (2011) *The Practice of Everyday Life.* Translated by SF Rendall. Berkeley: University of California Press.

Gately I (2014) *Rush Hour: How 500 Million Commuters Survive the Daily Journey to Work.* London: Head of Zeus.

Glick Schiller N and Salazar NB (2013) Regimes of Mobility across the Globe. *Journal of Ethnic and Migration Studies* 39(2): 183–200.

Goffman E (1959) *The Presentation of Self in Everyday Life.* New York: Anchor.

Hannerz U (1980) *Exploring the City: Inquiries Toward an Urban Anthropology.* New York: Columbia University Press.

Harvey D (2006) Space as a Keyword. In: Castree N and Gregory D (eds) *David Harvey: A Critical Reader.* Oxford: Blackwell, pp. 270–294.

Kaufmann V, Bergman MM and Joye D (2004) Motility: Mobility as Capital. *International Journal of Urban and Regional Research* 28(4): 745–756.

Kav LaOved, Worker's Hotline (2012) Palestinian Workers. Available at: www.kavlaoved.org.il/en/areasofactivity/palestinianworkers/ (accessed 5 July 2015).

Kelly T (2006) *Law, Violence and Sovereignty Among West Bank Palestinians.* Cambridge: Cambridge University Press.

Kothari U (2013) Geographies and Histories of Unfreedom: Indentured Labourers and Contract Workers in Mauritius. *The Journal of Development Studies* 49(8): 1042–1057.

Lefebvre H (1991) *The Production of Space.* Translated by D Nicholson-Smith. Oxford: Blackwell.

Lefebvre H (ed.) (2013) *Rhythmanalysis: Space, Time and Everyday Life.* Translated by S Elden and G Moore. London: Bloomsbury Academic.

Lefebvre H and Régulier C (2013) Attempt at the Rhythmanalysis of Mediterranean Cities. In: Lefebvre H (ed.) *Rhythmanalysis: Space, Time and Everyday Life.* Translated by S Elden and G Moore. London: Bloomsbury Academic, pp. 85–100.

LeVine M (2005) *Overthrowing Geography: Jaffa, Tel Aviv, and the Struggle for Palestine, 1880–1948.* Berkeley: University of California Press.

Maltz J (2014) *The Gaza Effect: In Shaky Economy, Arab Businesses Hit Hardest.* Haaretz, 7 August. Available at: http://www.haaretz.com/news/national/.premium-1.609299 (accessed 5 July 2015).

Orpaz I (2015) *Arabs Taking Their Place in Startup Nation.* Haaretz, 24 January. Available at: http://www.haaretz.com/business/.premium-1.570280 (accessed 5 July 2015).

Ortner SB (2006) *Anthropology and Social Theory: Culture, Power, and the Acting Subject.* Durham, NC: Duke University Press.

Pappé I (2011) *The Forgotten Palestinians: A History of the Palestinians in Israel.* New Haven: Yale University Press.

Portugali J (1993) *Implicate Relations: Society and Space in the Israeli-Palestinian Conflict.* Dordrecht: Kluwer Academic Publishers.

Rabinowitz D and Abu-Baker K (2005) *Coffins on Our Shoulders: The Experience of the Palestinian Citizens of Israel.* Berkeley: University of California Press.

Rabinowitz D and Monterescu D (2007) Introduction: The Transformation of Urban Mix in Palestine/Israel in the Modern Era. In: Rabinowitz D and Monterescu D (eds) *Mixed Towns, Trapped Communities: Historical Narratives, Spatial Dynamics, Gender Relations and Cultural Encounters in Palestinian-Israeli Towns (Re-materialising Cultural Geography).* Aldershot: Ashgate, pp. 1–32.

Robinson S (2013) *Citizen Strangers: Palestinians and the Birth of Israel's Liberal Settler State.* Stanford: Stanford University Press.

Roy S (2007) *Failing Peace: Gaza and the Palestinian-Israeli Conflict.* London: Pluto Press.

Salazar NB and Smart A (2011) Anthropological Takes on (Im)Mobility. *Identities: Global Studies in Culture and Power* 18(6): i–ix.

Saunders D (2011) *Arrival City: How the Largest Migration in History Is Reshaping Our World.* New York: Vintage.

Shamir R (2005) Without Borders? Notes on Globalization as a Mobility Regime. *Sociological Theory* 23(2): 197–217.

Urry J (2007) *Mobilities.* Cambridge: Polity Press.

Wacquant L (2008) *Urban Outcasts: A Comparative Sociology of Advanced Marginality.* Cambridge: Polity Press.

Wacquant L (2011) A Janus-Faced Institution of Ethnoracial Closure: A Sociological Specification of the Ghetto. In: Hutchinson R and Haynes BD (eds) *The*

Ghetto. Contemporary Global Issues and Controversies. Boulder: Westview Press, pp. 1–32.

Wesley DA (2013) *State Practices and Zionist Images: Shaping Economic Development in Arab Towns in Israel.* Oxford and New York: Berhand Publishers.

Yiftachel O (2001) Centralized Power and Divided Space: Fractured Regions in the Israeli Ethnocracy. *GeoJournal* 53(3): 283–293.

Yiftachel O (2002) The Shrinking Space of Citizenship: Ethnocratic Politics in Israel. *Middle East Report* 223: 38–45.

Yiftachel O (2006) *Ethnocracy: Land and Identity Politics in Israel/Palestine.* Philadelphia: University of Pennsylvania Press.

10
Reframing the Olympic Games: Uncovering New Spatial Stories of (De)securitization

Faye Donnelly

Introduction

This chapter reframes the Olympic Games as a space of contestation rather than a fixed place of peace. Initially this objective sounds ludicrous given that this mega-event was deliberately designed as a pathway to transcend conflict and instigate peace. In ancient times the proclamation of a sacred truce during the four Panhellenic Games called on all warring parties to halt their battles before, during and after the competition to ensure the total safety of athletes and spectators (Georgiadis and Syrigos 2009). While no truce is formally declared in modern epochs, the expectation that peace will prevail during the games is inscribed in the Olympic Charter (IOC 1894[2014]) and is avowedly championed by the International Olympic Committee (IOC) and Olympic Movement. Such sentiments were echoed by the UN general secretary, Ban Ki-Moon, during the preparations for the Sochi 2014 Winter Olympics. As he noted,

> the Olympic Truce is rooted in the hope that if people and nations can put aside their differences for one day, they can build on that to establish more lasting ceasefires and find paths towards durable peace, prosperity and human rights. For these next few weeks, may the torch of the Olympic and Paralympic Games in Sochi remind us what is possible when nations unite.
>
> (Ban 2014)

The contemporary ambitions of the Olympic Games to create peace go beyond the ancient truce: 'The goal of the Olympic Movement is to contribute to building a peaceful and better world by educating youth people through sport practised in accordance with Olympism and its values' (IOC 1894[2014]: 15). Elsewhere the Olympic Charter openly declares that 'the practice of sport is a human right' and that,

> The enjoyment of the rights and freedoms set forth in this Olympic Charter shall be secured without discrimination of any kind, such as race, colour, sex, sexual orientation, language, religion, political or other opinion, national or social origin, property, birth or other status.
>
> (IOC 1894[2014]: 11–12)

Through statements such as these, agents (re)constitute and reinforce the Olympics as a fixed place of peace.

Yet, to stop here in the story largely misses the point. Despite repeated attempts to reify the Olympics as a fixed place of peace, this ending is never guaranteed. By extension, this place is never totally secure. Taking these observations seriously means that the long-standing spatial story of the Olympics is constantly open to contestation and even rupture. These possibilities cannot be omitted. As history clearly demonstrates, the Olympics are not immune from boycotts, protest, violence and even death (Bajoria 2008; Cottrell and Nelson 2011; Hill 1996; Lenskyj 2000; Karamichas 2013; Schaus and Wenn 2007).

On a deeper level, this chapter argues that if we blindly accept storylines that axiomatically connect the Olympics and peace, we may obscure the heterogeneity of spaces and stories in play. With remarkable constancy another storyline is characterizing the Olympics according to a sequence of 'on your marks, get set, securitise' (Coaffee and Rogers 2008; Hellberg 2014; Lenskyj 2000, 2002; Toohey and Taylor 2012). This alternative story has many facets. For some the Olympics are considered to be securitized on the grounds that host states are increasingly vulnerable to an array of catastrophic attacks (Falcous and Silk 2006; Giulianotti and Klauser 2012; Toohey and Taylor 2008). The risk of agents hijacking this venue to make a statement is a real security concern for any host nation. As Hedenskog reveals, 'the power of attraction' created by the games

> also makes them the prime targets for individuals or groups that want to create attention around political or secessionist demands, as well as opportunity to strike against central authority at a time when it will

be vulnerable and highly sensitive to any assaults against its security. (2014: 180)

A directly interrelated line of argument is that the games have been securitized through the use of extensive surveillance policies (Bennett and Haggerty 2011; Boyle and Haggerty 2009). A common grievance raised here is that the implementation of such 'heightened security measures' in the build-up, duration and aftermath of each Olympics erodes rather than protects civil liberties and freedoms (Haggerty and Ericson 2000; Hassan 2014; Sugden 2012; Vlcek 2007).

However, suggesting that the games are securitized may not get us very far in reframing them as a space of contestation rather than a fixed place of peace. At best, securitization provides a useful entry point for understanding how the Olympics become a space of political protest and existential threat. It also helps us to examine how security is spoken in this context and, in turn, how different agents undertake multiple (de)securitizing moves. Nevertheless, as spelled out later, it is important to be careful about how securitized stories are told and believed.

By reframing the Olympics as a space of contestation, I hope to provide a timely reminder that they do not have to be securitized. On the contrary, there is no clear way in which securitization and this mega-event are mutually compatible. Returning to the principles inscribed in the Olympic Charter, this mega-event should be immune from any type of politicization or securitization. Making this plain, the latter explicitly notes:

> Recognising that sport occurs within the framework of society, sports organisations within the Olympic Movement shall have the rights and obligations of autonomy, which include freely establishing and controlling the rules of sport, determining the structure and governance of their organisations, enjoying the right of elections free from any outside influence and the responsibility for ensuring that principles of good governance be applied.
>
> (IOC 1894[2014]: 11)

Before proceeding it is necessary to clarify the following points. First, this chapter in no way rejects the fact that intimate relationships exist between peace and the Olympics. Undoubtedly the games have been, and continue to be, used as a way to improve intra-, inter- and supranational relations. Decisions to fly the UN flag at all competition sites of the Olympic Games in 1998 (UN General Assembly 2002), and to bring the Olympic flame to the UN for the first time in 2004, exemplify the common aspiration of both organizations to create international peace. Equally, this chapter does not argue that the games cannot desecuritize antagonistic disputes. Such a claim would be simplistic given that many countries have sought to create new agendas for peace and to ameliorate old hostilities under the Olympic banner. A contemporary example here is the inclusion of Dublin, the capital of the Republic of Ireland, as a destination for the London 2012 Olympic torch relay (Addley 2012). Speaking at the opening ceremony of the latter event, Sebastian Coe[1] stated: 'in 1948, shortly after the Second World War, my predecessor stood where I am today and made the first tentative steps in turning the world from war to sport' (cited in Kelso 2012). However, accepting that the Olympics have the potential to turn 'the world from war to sport' is not difficult given its stated purpose. The puzzle confronting us is how it is possible for the games to create securitization rather than peace? This is the research puzzle I am going to address. However, to reiterate the goal here is not to assume that the Olympics are securitized but to take a step back to ask how this outcome is becoming possible in this particular context.

The remainder of the chapter is structured as follows. Drawing on the work of Michel de Certeau, the first section introduces the spatial perspective adopted throughout. Next I analyse the Copenhagen School's securitization framework and a number of its critiques. The goal here is not to resolve disputes over how the original securitization framework can be advanced. Instead I attempt to bridge de Certeau's concept of 'spatial stories' with securitization to create new lines of inquiry. The third section shifts our attention to the empirical analysis. There I trace some of the core 'spatial stories' constructed to frame the 2014 Winter Olympics as a fixed place of peace on the one hand and securitized space on the other. Through my analysis it becomes clear that these two different stories are in dialogue with each other. It is also highlighted that there is a precarious balance between them. By way of conclusion, the final section relays the promises of reframing the games as a space of contestation.

Methodology

To conceptualize the 2014 Winter Olympics as a space of contestation I set out to examine the complicated semantic nexuses surrounding this mega-event. However, I was quickly confronted with the dilemma of 'information overload'. During the 17 days of this competition period alone, the world was inundated with multiple speech acts, ranging from official IOC press statements, to speeches made by national governments, to spectacular opening and closing ceremonies, to interviews with athletes, to boisterous protest rallies. Moreover, each of these utterances attracted extensive media coverage that was disseminated instantaneously across the globe through various communicative platforms. As Gerhard Heiberg, chairman of the IOC Marketing Commission, emphasized,

> from a broadcasting perspective, these Games broke new ground by offering more coverage on more channels and platforms than ever before, with the amount of digital coverage of Sochi 2014 exceeding traditional television broadcasts for the first time in Olympic history.
> (IOC 2014: 7)

Confronted with an inundation of speech acts, this chapter limits itself to considerations of how security was spoken by key actors before, during and after the 2014 Winter Olympics. From my analysis, the IOC and official Russian spokespeople emerged as the main speakers for peace and security in this context. This observation is somewhat counterintuitive given the amount of venomous scrutiny these actors came under over how they were running the games. However, as illustrated, a variety of voices emerged to challenge the dominant spatial story that was advocated by Russia and the IOC.

Aware that discourse analysis is not a homogenous methodological approach, it is necessary to mention that I concentrated on verbal and written texts of the dominant speakers at the expense of non-verbal or visual genres (Milliken 1999). This decision is justified since the primary goal here is to explore what types of (de)securitizing moves were undertaken to (re)frame the 2014 Winter Olympics as a secure and insecure space. A crucial limitation of the analysis presented is that I do not speak fluent Russian. Consequently, I relied on English-based sources and translations to engage with the speech acts uttered by Russian spokespeople in this context. However, it was possible for me to access the majority of texts relevant to this study because the official languages of the IOC are English and French (IOC 2015). Nevertheless, the question of how the analysis presented here travels to other spatial and linguistic milieus should be a future line of inquiry.

Theorizing spatial perspectives

Today it is habitual for scholars to speak of a 'spatial turn' within international relations and other disciplines (Hubbard and Kitchin 2011; Tally 2013; Warf and Arias 2014). However, despite the ascendancy of this 'turn', space remains notoriously difficult to pin down in theory and in practice. To reframe the Olympic Games as a space of contestation rather than a fixed place of peace, this section outlines the writing of Michel de Certeau. My reasons are three-fold. First, de Certeau provides a lens to extrapolate the interrelationships and differences between place and space. Second, foregrounding his concept of spatial stories enriches our analytical toolbox for uncovering how actors attempt to impose order on fluid spaces. Third, taking his ideas on board shows that different (de)securitizing moves were attempted during the 2014 Winter Olympics.

Space and place

In essence, de Certeau studies the multiplicity of space and its constant mobility. According to him, 'space is composed of intersections of mobile elements. It is in a sense actuated by the ensemble of movements deployed within it' (de Certeau 1988: 117). An important point foregrounded in this quotation is that space is constituted by, and constitutive of, multidimensional spheres of interaction. In short, it is not fixed. On the contrary, de Certeau portrays space as a tapestry of complexities that can never be reduced to, or contained by, static boundaries. Rejecting any kind of reductive delineations, he conceptualizes space as a 'practice' (ibid.). Considered from this angle, it is not possible for actors to step outside space to understand it in a detached fashion. On the contrary, spatial practices are lived in and experienced through everyday settings, encounters and 'pedestrian speech acts' (1988: 97; also see de Certeau 2014).

Although de Certeau depicts spatial practices as fluid and mobile parameters of action, he does not portray them as benign or emancipatory projects. Instead, spatial practices can transcend limits but they can also be reduced to them. It is in the latter capacity that de Certeau calls on his readers to become critically aware of how spatial practices can normalize power relations to an illusionary inertia (1988: 125–126, 201). Indeed, as a critical project, his work seeks to expose the banal rituals of inclusion and exclusion that are constituted and enforced by certain orders.

Another defining feature of de Certeau's theorization of space is its distinguishability from place. He makes this distinction on the grounds that the latter 'implies an indication of stability', whereas the former takes into consideration 'vectors of direction, velocities and time variables' (1988: 117). Going further, he notes:

> in relation to place, space is like the word when it is spoken, that is, when it is caught in the ambiguity of an actualisation, transformed into a term dependent on many different conventions, situated as the act of a present (or of a time), and modified by the transformations caused by successive contexts. In contradistinction to the place, it has thus none of the univocity or stability of a 'proper'.
>
> (1988: 117)

An outstanding observation here is that spaces and places both constitute and enforce a certain order.

Spatial stories

Throughout his writings, de Certeau reflexively problematizes the binary he draws between space and place. He does this by suggesting that space can transform into place and vice versa. Here, 'spatial stories' are deemed to be vitally important. Ultimately, de Certeau claims 'spatial stories' facilitate transitions between spaces and places. To be exact, stories 'transverse and organise place; they select and link them together; they make sentences and itineraries out of them. They are spatial trajectories.' From this baseline he further argues that 'every story is a travel story – a spatial practice' (1988: 115).

What stands out in de Certeau's discussion of spatial stories is that they are not simply told. As spatial practices, they are enacted. By the same token, spatial stories are intimately interwoven into how spaces and places are constructed, inhabited and embodied. Reciprocally, where stories are disappearing or deprived of narrations, there is a loss of space (1988: 123). Another salient observation de Certeau makes here is that spatial stories always remain polyvalent and thus can be interconnected with other stories. Taking this line of argument seriously ensures that opposing stories can be combined while always remaining separate and unfinished (1988: 119).

Against this backdrop, the remaining sections incorporate de Certeau's ideas to reframe the games as a space of contestation rather than a fixed place of peace. Doing this also paints a more holistic picture of how the Olympics can be securitized in theory and in practice.

Security speech acts and securitization

According to the Copenhagen School, security does not exist *a priori* or unfold in any predictable way. Rather, it contends, 'security is a self-referential practice, because it is in this practice that the issue becomes a security issue – not necessarily because a real existential threat exists but because the issue is presented as such a threat' (Buzan et al. 1998: 24). Inspired by Austin's speech act theory, the Copenhagen School claims that speaking security does something. By uttering a security speech act, agents undertake a 'securitizing move' that identifies a referent object as being existentially threatened by a certain issue. In the most general terms, securitizing speech acts and moves signify attempts to legitimate the use of extraordinary measures to deal with said security threat. In effect, this requires agents to shift issues outside of the rules that 'would otherwise bind' (Buzan et al. 1998: 24).

It is imperative to note that neither security speech acts nor securitizing move(s) equate to securitization. The latter requires audience acceptance. For, 'if no sign of such acceptance exists, we can talk only of a securitizing move, not of an object actually being securitized (Buzan et al. 1998: 25). By making audience acceptance a paramount prerequisite for securitization, the Copenhagen School reinforces the fact that speaking security is not the ideal choice or a banal action. Instead 'it is always a political choice to securitise or accept a securitisation' (1998: 29). Building from this platform the Copenhagen School goes so far as to contend that securitization is a failure (1998), and that this failure emanates from the inability of agents to resolve issues within the normal political sphere.

It is in this vein that desecuritization becomes pivotal. Indeed, Vuori presents 'de-securitisation as a counter-strategy or-move to securitisation' (Vuori 2011: 191). Within the Copenhagen School framework, this concept is defined as 'the shifting of issues out of emergency mode and into the normal bargaining processes of the political sphere', thus moving them out of a 'threat-defense sequence and into the ordinary public sphere' (Buzan et al. 1998: 4, 29; Wæver 1995). Typically, desecuritization is presented as the preferable long-range option 'since it means not to have issue phrased as "threat against which we have countermeasures" but to move them out this threat-defense sequence and into the ordinary public sphere' (Buzan et al. 1998: 29).

Two points should be taken away. First, 'securitization is a process that unfolds as agents start and stop speaking security' (Donnelly 2013: 45).

Second, the Copenhagen School is largely averse to the materialization of securitized space.

Securitization and 'spatial stories'

To date, the Copenhagen School's securitization framework has been the subject of tremendous debate. Some sceptics maintain that this concept is not emancipatory and as such should be abandoned altogether (Booth 1991, 2007; Dalby 2009). Other scholars engaging with the original framework favour modification as opposed to total abandonment. A large part of what is going on in the latter debates touches on contextual considerations (cf. Balzacq 2005, 2011, 2015; Ilgit and Klotz 2014; McDonald 2008; Salter 2008; Stritzel 2007). Specifically, Balzacq has advanced a more 'sociological approach' to securitization as a way to solidify '1) the centrality of audiences; 2) the co-dependency of agency and context; 3) the structuring force of the dispotif, that is a constellation of practices and tools' (Balzacq 2011: 3). Concurrently, his alternative approach calls for a richer appreciation of how 'securitisation occurs in a field of struggles' (2011: 5).

Amid the burgeoning securitization debates, few attempts have been made to incorporate de Certeau's conceptualization of 'spatial stories' into the fold. This absence is particularly notable for three reasons. First, de Certeau may provide the Copenhagen School with a more robust reply to those who suggest that this process is not emancipatory. Although traces of this argument find expression in the Copenhagen School's claim that desecuritization should always be the ideal, bringing de Certeau in would reinforce the fact that spaces and places are never benign. Second, his emphasis on conceptualizing space as polyvalent would enable scholars to show that 'context' matters in securitization studies. To recall, de Certeau contends that 'space is composed of intersections of mobile elements ... actuated by the ensemble of movements deployed within it' (de Certeau 1988: 117). Finally, I would like to add another contribution by suggesting that securitization can be easily reframed as an ongoing spatial story. The innovation here could be important for changing different scholarly attitudes towards this approach. What is then distinctive about the Copenhagen School's securitization framework is the explicit and implicit spatial perspectives it contains about the (re)telling of security and (de)securitized stories. Of course, rooting securitization in different spatial stories could make it more relevant to those exploring how security is spoken in banal, everyday local and urbanized environments (Becker and Muller 2013; Huysmans 2011). Beneficially, de

Certeau shows that there are countless stories about space and place for (de)securitization.

The Sochi 2014 Winter Olympics: A space of contestation, peace or securitization?

On 4 July 2007 the president of the IOC, Jacques Rogge, announced Sochi as the host city for the 2014 Winter Olympics, to be held in 2014 (Olympic.org 2014). The games were organized in two clusters: a coastal cluster for ice events in Sochi, and a mountain cluster located in the Krasnaya Polyana Mountains.

For many the 2014 Winter Olympics signifies a carefully choreographed sequence of 'political mythmaking' that showcased Russia as a powerful and unified country (Müller 2011; Persson and Petersson 2014; Petersson and Vamling 2013). As Richmond claimed, 'from the start Moscow hailed the selection of Sochi as the host to the Winter Olympics as a multifaceted victory for the Russian people' (Richmond 2013: 203). Affirming this point, Dmitry Chernyshenko, president of the Sochi organizing committee, proclaimed: 'The Sochi 2014 Winter Olympics showed the new face of Russia to the world. We constructed – from scratch – state of the art sport venues that set the perfect stage for the performance of a lifetime' (IOC 2014: 8).

The purpose of this section is to highlight how the long-standing spatial story of the Olympics as a fixed place of peace was reproduced and then contested in this context. As will be shown, there was an extreme disjuncture between views of what was thought to be required for peace to prevail at these games.

Setting the scene: Spatial stories of peace

Within the parameters of the spatial story espoused by the IOC and Russian officials, the 2014 Winter Olympics were clearly denoted a fixed place of peace. From the outset it was declared: 'Russia is determined to complete all the projects at the highest level in due time – and of course in full cooperation with the IOC. The Games have the full support of the Russian government' (Russian Federation Deputy Prime Minister Alexander Zhukov, cited in IOC 2010).

This line of argument finds expression in several interrelated narrative strands. One area where it manifested itself most interestingly was in the considerable effort Russia devoted to ensuring that the 2014 Winter Olympics were safe and secure. The eradication of any types of insecurity thus became a top priority. Indeed, the whole semantics framing the 2014 Winter Olympics as a fixed place of peace were very frequently rationalized through concerns about safety. Under the slogan 'safest in history', Russia choreographed an extensive array of security measures (Cyveillance 2014: 2). For example, in August 2013 the Kremlin issued a presidential decree that severely restricted access to and from the Black Sea resort. This resulted in the demarcation of a vast 'forbidden zone' as well as a number of 'controlled zones' (BBC 2013a). Enveloped in the decrees was also the prohibition of 'protests, meeting, rallies demonstrations, marches and pickets from 7 January 2014 to 21 March 2014' (Cyveillance 2014: 3). Notably, these measures were justified by the Russian authorities and the IOC in the name of safety and security measures. As Chernyshenko memorably stated, Sochi was the 'most secure venue on the planet' (cited in Wilson 2014).

'Transcending the limits'

Irrespective of Russia and the IOC putting the long-standing spatial story of the Olympics as a fixed place of peace into practice, their speech acts quickly became a nodal point for contestation and opposition long before the opening ceremony took place in Sochi on 7 February 2014. Immediate concerns were raised about the suitability of Sochi as a venue for the 2014 Winter Olympics given the subtropical climate of the Black Sea summer resort (*The Economist* 2013). More seriously, many activists condemned 'the forced evictions of some 2,000 families to make way for the Olympic venues and infrastructure', alongside 'migrant workers abuse, environmental degradation, crackdowns on civil society and the press' (HRW 2014). Charges of corruption at what the Russian president, Vladimir Putin, described as 'the biggest building site on the planet' also tainted the event (Lenskyj 2014: 3; Putin 2014; also see Gibson 2013).

At a more regional level, attempts by Russia to frame the 2014 Winter Olympics as fixed place of peace was complicated by 'a centuries-old row it has with the disposed Circassian nation' (Zhemukhov 2010). Here the actual sites of the Olympic stadia aroused powerful memories and strong emotions (Zhemukhov and Orrtung 2014). As one commentator noted, 'Circassians are saying don't play games with the memory of our ancestors and since the Games represent peace and dialogue between civilisations, don't hold this event on the site of a genocide' (cited in Sahin 2013).

The ability of Russia to host secure games in Sochi was also questioned on the grounds that 'this region has been shaken by ethnic conflicts and has experienced two Chechen wars. It is also the part of Russia over which Islamic radicals have the strongest hold' (Malashenko 2013).

Fear of possible terrorist attacks by regional actors were fuelled when Doku Umarov, head of the Russian group Imarat Kavkaz, released a video dated June 2013 that called on his rebels to disrupt the upcoming Winter Olympics (Edler 2013). The suicide bomber attacks in Volgograd that killed 34 people in December 2013 did little to dampen suspicion that 'Sochi itself would be turned into a tantalizing target for Islamic terrorists' (Myers 2014).

Vociferous concerns were also expressed about the use of an enhanced version of SORM, Russia's domestic electronic eavesdropping system (Galeotti 2013). In the face of what was described as 'Russia's surveillance state' (Soldatov and Borogan 2013), the US Department of Homeland Security issued a series of 'security tips' to those travelling to Sochi and those watching the games abroad. Apart from raising the possibility of 'hacktivists', these tips also noted that

Russia has a national system of lawful interception of all electronic communications. The System of Operative-Investigative Measures, or SORM, legally allows the Russian FSB to monitor, intercept, and block any communication sent electronically (i.e. cell phone or landline calls, internet traffic, etc.).

(US Department of Homeland Security 2014)

Storylines that equated the 2014 Winter Olympics with a fixed place of peace were seriously strained in relation to the anti-gay propaganda laws that the Russian Government passed in the period leading up to the Olympics. Several developments in 2013 raised fears of an 'open season' on lesbian, gay, bisexual and transgender (LGBT) people, 'including the torture and murder of a young gay man, a violent attack on a trans woman, and a Russian member of parliament's call for public whipping of gay men' (Lenskyj 2014: 4). Adding weight to these concerns, HRW reported: 'this law openly discriminates against LGBT people, legitimises anti-LGBT violence, and seeks to erase LGBT people from the country's public life' (HRW 2014). Tensions over the rights of LGBT individuals in Russia culminated in the branding of the 2014 Winter Olympics as the 'Gay Olympics' (Friedman 2014). In tandem, this issue was politicized to the point of overt calls for governments to boycott the games. One example of moves in this direction came from Nadezhda Tolokonnikova, a member of the Russian punk band Pussy Riot, shortly after she had been released from jail (BBC News 2013b). Fry also warned that inaction on this matter would ensure that the Olympic 'five rings would finally be forever smeared, besmirched and ruined in the eyes of the civilized world' (2013). Likewise, President Barack Obama included three openly gay athletes in the US Olympic delegation while refraining from attending himself (Office of the Press Secretary 2014).

A space of contestation

Against the tension-laden backdrops sketched above, the 2014 Winter Olympics appear to be anything but a fixed place of peace. Instead, this mega-event can easily be conceptualized as a securitized space. Which of these two radically different spatial stories is correct? Can they be reconciled? It is precisely at this point that the promise of reframing the 2014 Winter Olympics as a space of contestation shines through. From this viewpoint it is clear that this mega-event became a polyvalent space in which securitizing and desecuritizing moves were enacted simultaneously. Thus the two stories do not exist in splendid isolation. More frequently than not they were enacted in the same places and in the same spaces to speak security and peace in different ways.

Returning to de Certeau's concept of spatial stories also exposes how different actors sought to promote and prioritize certain spatial stories over others. Something that is often overlooked in discussions about the 2014 Winter Olympics is that it is possible to (re)tell either story from a different perspective. To exemplify, while the acts of protest mounted in the build-up to, and during, the event are largely framed as desecuritizing moves, they can also be viewed as attempted securitizing moves against the Russian Government and even the IOC. As mentioned, both of these actors were in different respects categorized as posing an existential threat to various groups, ranging from LGBT communities, to Circassians, to migrant worker, to spectators. Leaving aside whether these securitizing claims are true, the point being made here is that these (de)securitized stories have a counternarrative and many competing strands of meaning embedded in them.

Interestingly, my analysis has also hinted that deliberate desecuritizing moves are interwoven with the allegedly securitized policies of the Russian Government. In response to rising concerns about the new propaganda laws he had passed, for instance, President Putin assured the international community that LGBT athletes or visitors would not be subject to any kind of intimidation during their stay in Russia. Instead he stressed that they 'can feel relaxed and calm, but leave children alone please' (cited in Walker 2014). It is also possible to suggest that the Russian Government's decision to retract the absolute ban on public protests and demonstrations that was originally legitimated under the 2013 presidential decree constitutes a desecuritizing move. The latter 'move' was particularly welcomed by the IOC's executive board on the grounds that 'the people will now have an opportunity to express their views and freely demonstrate their opinions in Sochi' (IOC 2013a).

Conceptualizing the 2014 Winter Olympics in Sochi solely as a securitized story also overlooks how the IOC reaffirmed the games a fixed place of peace. Addressing the rising concerns over Russia's anti-gay legislation, for instance, it stressed that it had 'received assurances from the highest level of government in Russia that the legislation will not affect those attending or taking part in the Games' (IOC 2013b). In conjunction, the IOC president, Thomas Bach, took a firm stance against the games being used as a space of protest. Speaking at the opening ceremony, he noted: 'we are grateful to those who respect the fact that sport can only contribute to development and peace if it is not used as a stage for political dissent or for trying to score points in internal or external political contests' (Bach 2014). Here again, the story being told conforms more to a desecuritizing logic than to a securitized one.

Markedly, the plots, characters and endings of the stories constructed about the 2014 Winter Olympics vary dramatically. An important improvement in the spatial stories told about the event would be to reframe them as competing but also contested discursive frames and (de)securitizing strands. This would unmask the presence of multiple spatial enactments rather than simply pitting one story against the other.

Conclusions

Relationships between the Olympic Games, peace and security are complicated, multilayered and changing. However, there is a tendency for contemporary accounts of the Olympics to coalesce around two dominant stories. Proponents of the contend that this sporting occasion signifies a fixed place of peace. This narrative has a long legacy. Nevertheless, an antithetical supposition has now emerged which suggests that the Olympics are securitized.

Drawing on the work of Michel de Certeau, this chapter has argued that neither story is fully complete. Instead, each is better understood as a segment of a much larger totality of enacted spatial practices. As de Certeau reaffirms, no place is ever fixed and thus no space is ever fully secure. Conversely, spatial practices and stories are always multifaceted and mobile. This revelation entails a number of profound implications. If taken seriously, de Certeau's ideas can convert the spatial stories currently being told about and through the games. First and foremost his ideas disturb the boundaries of the dominant spatial story of the Olympics – namely that they are a fixed place of peace. As illustrated, the manifold contestations that erupted in relation to the 2014 Sochi Winter Olympics demonstrate that the IOC must constantly put the meanings of its charter into use no matter what place or space it is operating in (Williamson 2014).

Another benefit of drawing on de Certeau is that it allows us to reframe the Olympics as a space in which securitization can occur, rather than a securitized space *de facto*. This may seem like a trivial differentiation. However, conceptualizing the games in this way offers a much richer picture of the complexities at play. At first glance the 2014 Winter Olympics appear to confirm that this mega-event was securitized. Clearly extensive arrays of security measures were installed to monitor anyone attending and watching this particular sports spectacle. Arguably, such measures indicate that Russia felt existentially threatened by an array of actors and referent objects located inside and outside its national borders. However, returning to de Certeau, the limits of this securitized story can be transgressed. Indeed, as shown, there is evidence to suggest that the Russian Government and the IOC sought to continuously desecuritize the 2014 Winter Olympics and reaffirm them as a fixed place of peace.

Thus the types of spatial stories that are told about and enacted in the Olympics have enormous ramifications. In many ways, the mega-event we know today is becoming a product of spatial stories of securitization. This scenario is bleak because it suggests that the Olympics have been transformed from a fixed place of peace to a space where spectators and athletes alike are existentially threatened. Hence a negative consequence of securitizing the games is that this space comes to be framed as a perilous zone where people are never safe. It remains to be ascertained which spatial story – peace or securitization – will prevail or how either will end. If we follow de Certeau, the goal must be to resist any fixed ending.

Note

1. Chairman of the London 2012 Olympic Games organizing committee.

References

Addley E (2012) Olympic Torch Route, Day 19: A Tale of Two Cities, One Flame and Jedward. *The Guardian*, 6 June. Available at: www.theguardian.com/

sport/2012/jun/06/olympic-torch-relay-dublin-jedward (accessed 23 September 2014).

Bajoria J (2008) Politics and the Olympics. *Council of Foreign Relations*, 28 May. Available at: www.cfr.org/culture-and-foreign-policy/politics-olympics/p16366 (accessed 19 September 2014).

Bach T (2014) *Speech on the Occasion of Opening Ceremony – 126th Session, Sochi.* 4 February Available at: www.olympic.org/Documents/IOC_Executive_Boards _and_Sessions/IOC_Sessions/126_Session_Sochi/Sochi_2014_IOC_President _Bach_Speech_Session_Opening.pdf (accessed 28 April 2015).

Balzacq T (2005) The Three Faces of Securitization: Political Agency, Audience and Context. *European Journal of International Relations* 11(2): 171–201.

Balzacq T (ed.) (2011) *Securitization Theory: How Security Problems Emerge and Dissolve.* Abingdon: Routledge.

Balzacq T (2011) A Theory of Securitization: Origins, Core Assumptions, and Variants. In: Balzacq T (eds) *Securitization Theory: How Security Problems Emerge and Dissolve.* Abingdon: Routledge, pp. 1–30.

Balzacq T (ed.) (2015) *Contesting Security: Strategies and Logics.* Abingdon: Routledge.

Ban K-M (2014) *Secretary-General Calls on Combatants Everywhere to Respect Olympic Truce, Evoking Hope for More Lasting Ceasefires, Durable Peace, Prosperity.* SG/SM/15625, 31 January. Available at: www.un.org/News/Press/docs/2014/ sgsm15625.doc.htm (accessed 20 September 2014).

BBC News (2013a) Russia Bans Public Protest at 2014 Sochi Winter Olympics, 23 August: Available at: www.bbc.co.uk/news/world-europe-23819104 (accessed 28 April 2015).

BBC News (2013b) *Pussy Riot Member Urges Russia Olympics Boycott.* 23 December. Available at: www.bbc.co.uk/news/world-europe-25496255 (accessed 14 November 2014).

Becker A and Müller MM (2013) The Securitization of Urban Space and the 'Rescue' of Downtown Mexico City: Vision and Practice. *Latin American Perspectives* 40(2): 77–94.

Bennett CK and Haggerty KD (eds) (2011) *Security Games: Surveillance and Control at Mega-Events.* Abingdon and New York: Routledge.

Boyle P and Haggerty KD (2009) Spectacular Security: Mega-events and the Security Complex. *Political Sociology* 3(3): 257–274.

Buzan B, Wæver O and de Wilde J (1998) *Security: A New Framework for Analysis.* Boulder: Lynne Rienner Publishers.

Booth K (1991) Security and Emancipation. *Review of International Studies* 17(4): 313–326.

Booth K (2007) *Theory of World Security.* Cambridge: Cambridge University Press.

Coaffee J and Rogers P (2008) Rebordering the City for New Security Challenges: From Counter Terrorism to Community Resilience. *Space and Polity* 12(1): 101–118.

Cottrell MP and Nelson T (2011) Not Just the Games? Power, Protest and Politics at the Olympics. *European Journal of International Relations* 17(4): 729–753.

Cyveillance (2014) Sochi 2014: XXII Winter Olympic Security Concerns. Available at: info.cyveillance.com/rs/cyveillanceinc/images/CYV-WP-Sochi2014 PhysicalCyberThreats.pdf (accessed 19 April 2015).

Dalby S (2009) *Security and Environmental Change.* Cambridge: Polity Press.

de Certeau M (1988) *The Practice of Everyday Life*. Translated by Steven Rendall. Berkeley: University of California Press.

de Certeau M (2014) Spatial Practices: Walking in the City (1984). In: Gieseking J, Marigold W, Katz C, Low S and Saegert S (eds) *The People, Place and Space Reader*. New York: Routledge, pp. 232–236.

Donnelly F (2013) *Securitization and the Iraq War: The Rules of Engagement in World Politics*. Abingdon: Routledge.

Edler M (2013) Russian Islamist Doku Umarov Calls for Attacks on 2014 Winter Olympics. *The Guardian*, 3 July. Available at: www.theguardian.com/world/2013/jul/03/russia-islamist-attack-olympics-sochi/print (accessed 4 November 2014).

Falcous M and Silk M (2006) Global Regimes, Local Agendas: Sport, Resistance and the Mediation of Dissent. *International Review for the Sociology of Sport* 41(3–4): 317–338.

Friedman U (2014) How Sochi Became the Gay Olympics. *The Atlantic*, 28 January. Available at: www.theatlantic.com/international/archive/2014/01/how-sochi-became-the-gay-olympics/283398/ (accessed 14 November 2014).

Fry S (2013) *An Open Letter to David Cameron and the IOC*. 7 August. Available at: www.stephenfry.uk/2013/08/07/an-open-letter-to-david-cameron-and-the-ioc/ (accessed 15 November 2014).

Galeotti M (2013) On Your Marks, Get Set... Intercept!. *Open Democracy*, 29 October. Available at: www.opendemocracy.net/od-russia/mark-galeotti/on-your-marks-get-set%E2%80%A6-intercept (accessed 3 November 2014).

Georgiadis K and Syrigos A (2009) *Olympic Truce: Sport as a Platform for Peace*. International Olympic Truce Centre. Available at: olympism4humanity.files.wordpress.com/2013/04/5091_olympictruce.pdf (accessed 28 May 2015).

Gibson O (2013) Sochi 2014: The Costliest Olympics Yet but Where Has All the Money Gone? *The Guardian*, 9 October. Available at: www.theguardian.com/sport/blog/2013/oct/09/sochi-2014-olympics-money-corruption (accessed 13 November 2014).

Giulianotti R and Klauser F (2012) Sport Mega-Events and 'Terrorism': A Critical Analysis. *International Review for the Sociology of Sport* 47(3): 307–323.

Haggerty KD and Ericson RV (2000) The Surveillant Assemblage. *British Journal of Sociology* 51(4): 605–622.

Hassan D (2014) Securing the Olympics: At What Price? *Sport in Society* 17(5): 628–639.

Hedenskog J (2014) The Terrorist Threat against Sochi 2014. In: Petersson B and Vamling K (eds) *The Sochi Predicament: Contexts, Characteristics and Challenges of the Olympic Winter Games in 2014*. Newcastle: Cambridge Scholars, pp. 180–197.

Hellberg U (2014) Securitization in the North Caucuses on the Eve of the Sochi Games. In: Petersson B and Vamling K (eds) *The Sochi Predicament: Contexts, Characteristics and Challenges of the Olympic Winter Games in 2014*. Cambridge: Cambridge Scholars, pp. 159–179.

Hill CR (1996) *Olympic Politics: Athens to Atlanta 1896–1996*. Manchester: Manchester University Press.

Hubbard P and Kitchin R (2011) *Key Thinkers on Space and Place*. 2nd edn. London: Sage.

Human Rights Watch (2014) *Russia: Anti-LGBT Law a Tool for Discrimination*. An Anniversary Assessment, 30 June: Available at: www.hrw.org/news/2014/06/ 29/russia-anti-lgbt-law-tool-discrimination (accessed 15 November 2014).

Huysmans J (2011) What Is in an Act? On Security Speech Acts and Little Security Nothings. *Security Dialogue* 42(4–5): 371–383.

Ilgit A and Klotz A (2014) How Far Does 'Societal Security' Travel? Securitisation in South African Immigration Policies. *Security Dialogue* 45(2): 137–155.

IOC (2010) *Bid Procedure for the Olympic Winter Games of 2014*: Reference Document: Research and Reference Olympic Studies Centre. Available at: www.olympic.org/assets/osc%20section/pdf/qr_7e.pdf (accessed 27 April 2015).

IOC (2013a) *First IOC Executive Board Meeting under President Bach* 10 December: Available at: www.olympic.org/news/first-ioc-executive-board-meeting-under-president-bach/218358 (accessed 28 April 2015).

IOC (2013b) *IOC Statement on Recent Russian Legislation* 31 July. Available at: www.olympic.org/news/ioc-statement-on-recent-russian-legislation/206969 (accessed 15 November 2014).

IOC (1894[2014]) Olympic Charter. Available at: www.olympic.org/Documents/ olympic_charter_en.pdf (accessed 27 April 2015).

IOC (2014) *Marketing Report: Sochi 2014*. Available at: www.olympic.org/ Documents/IOC_Marketing/Sochi_2014/LR_MktReport2014_all_Spreads.pdf (accessed 19 April 2015).

IOC (2015) *Fact Sheet: The Olympic Movement: Update*. Available at: www.olympic .org/Documents/Reference_documents_Factsheets/The_Olympic_Movement .pdf (accessed 19 April 2015).

Karamichas J (2013) *The Olympic Games and the Environment*. Basingstoke: Palgrave Macmillan.

Kelso P (2012): London 2012 Olympics: Torch Begins Its Journey to London Following Lighting Ceremony in Olympia. *The Telegraph*, 10 May. Available at: www.telegraph.co.uk/sport/olympics/torch-relay/9256484/London-2012 -Olympics-torch-begins-its-journey-to-London-following-lighting-ceremony -in-Olympia.html (accessed 15 November 2014).

Lenskyj HJ (2000) *Inside the Olympic Industry: Power, Politics and Activism*. New York: State University of New York Press.

Lenskyj HJ (2002) *The Best Olympics Ever? Social Impacts of Sydney 2000*. New York: State University of New York Press.

Lenskyj HJ (2014) *Sexual Diversity and the Sochi 2014 Olympics: No More Rainbows*. Basingstoke: Palgrave Macmillan.

McDonald M (2008) Securitization and the Construction of Security. *European Journal of International Relations* 14(4): 563–587.

Malashenko A (2013) Controversy and Concern over the Sochi Olympics. *Carnegie Endowment for International Peace*, 10 April. Available at: carnegieendowment.org/2013/04/10/controversy-and-concern-over-sochi -olympics/fyyh (accessed 3 November 2014).

Milliken J (1999) The Study of Discourse in International Relations: A Critique of Research Methods. *European Journal of International Relations* 5(2): 225–254.

Müller M (2011) State Dirigisme in Megaprojects: Governing the 2014 Winter Olympics in Sochi. *Environment and Planning A* 43(9): 2091–2108.

Myers SL (2014) An Olympics in the Shadow of a War Zone. *The New York Times*, 5 February: Available at: www.nytimes.com/2014/02/05/world/europe/an-olympics-in-the-shadow-of-a-war-zone.html (accessed 4 November 2014).

Office of the Press Secretary, The White House (2014) *President Obama Updates Presidential Delegation to the Opening Ceremony of the 2014 Olympic Winter Games*, 5 February. Available at: www.whitehouse.gov/the-press-office/2014/02/05/president-obama-updates-presidential-delegation-opening-ceremony-2014-ol (accessed 14 November 2014).

Olympic.org (2014) *2014 Host City Election*. Available at: www.olympic.org/content/the-ioc/bidding-for-the-games/past-bid-processes/2014-host-city-election/ (accessed 27 April 2014).

Persson E and Petersson B (2014) Political Mythmaking and the 2014 Winter Olympics in Sochi: Olympism and the Russian Great Power Myth. *East European Politics* 30(2): 192–209.

Petersson B and Vamling K (eds) (2013) *The Sochi Predicament: Contexts, Characteristics and Challenges of the Olympic Winter Games in 2014*. Newcastle: Cambridge Scholars.

Putin V (2014) *Interview to Russian and Foreign Media*, 19 January: Available at: eng.kremlin.ru/transcripts/6545 (accessed 15 November 2014).

Richmond W (2013) Preparation for the Sochi Olympics. In: Ware RB (ed.) *The Fire Below: How the Caucasus Shaped Russia*. London: Bloomsbury, pp. 203–224.

Salter Mark B (2008) Securitization and Desecuritization: A Dramaturgical Analysis of Canadian Air Transport Security Authority. *Journal of International Relations and Development* 11(4): 321–349.

Sahin D (2013) Winter of Discontent. *Al Jazeera*, 19 June. Available at: www.aljazeera.com/programmes/aljazeeraworld/2013/06/20136179431945292.html (accessed 25 April 2015).

Schaus GP and Wenn SR (eds) (2007) *Onwards to the Olympics: Historical Perspectives on the Olympic Games*. Waterloo: Wilfrid Laurier University Press.

Soldatov A and Borogan I (2013) Russia's Surveillance State. *World Policy Journal* 13(3): 357–383.

Stritzel H (2007) Towards a Theory of Securitization: Copenhagen and Beyond. *European Journal of International Relations* 13(3): 357–383.

Sugden J (2012) Watched by the Games: Surveillance and Security at the Olympics. *International Review for the Sociology of Sport* 47(3): 414–429.

Tally RT (2013) *Spatiality: The New Critical Idiom*. Abingdon: Routledge.

The Economist (2013) *The Sochi Olympics: Castles in the Sand*, 13 July. Available at: www.economist.com/news/europe/21581764-most-expensive-olympic-games-history-offer-rich-pickings-select-few-castles (accessed 13 November 2014).

Toohey K and Taylor T (2008) Mega Events, Fear and Risk: Terrorism at the Olympic Games. *Journal of Sport Management* 22: 451–469.

Toohey K and Taylor T (2012) Surveillance and Securitization: A Forgotten Sydney Olympic Legacy. *International Review for the Sociology of Sport* 47(3): 324–337.

UN General Assembly (2002) *Solemn Appeal Made by the President of the General Assembly on 25 January 2002 in Connection with the Observance of the Olympic Truce [A/595]*. Available at: www.un.org/documents/ga/docs/56/a56795.pdf (accessed 8 November 2014).

US Department of Homeland Security (2014) *Security Tips (ST14-001)*, 4 February. Available at: www.us-cert.gov/ncas/tips/ST14-001 (accessed 3 November 2014).

Vlcek W (2007) Surveillance to Combat Terrorist Financing in Europe: Whose Liberty, Whose Security? *European Security* 16(1): 99–119.

Vuori JA (2011) Religion Bites: Falungong, Securitization/desecuritization in the People's Republic of China. In: Thierry Balzacq (ed.) *Securitization Theory: How Security Problems Emerge and Dissolve.* Abingdon, Oxon: Routledge.

Walker S (2014) Vladimir Putin: Gay People at Winter Olympics Must 'Leave Children Alone'. *The Guardian,* 17 January. Available at: www.theguardian.com/world/2014/jan/17/vladimir-putin-gay-winter-olympics-children (accessed 15 November 2014).

Warf B and Arias S (2014) *The Spatial Turn: Interdisciplinary Perspectives.* Abingdon: Routledge.

Wæver Ole (1995) Securitization and Desecuritization. In: Ronnie D. Lipschutz (eds) *"On Security":* New York: Columbia University Press.

Williamson H (2014) The IOC Needs a New Olympic Playbook after Russia Winter Olympics. *Policy Review.* Available at: www.policyreview.eu/the-ioc-needs-a -new-olympic-playbook-after-russia-debacle/ (accessed 18 September 2014).

Wilson S (2014) Sochi Chief: City is World's 'Most Secure Venue'. *Associated Press,* 29 January. Available at: www.washingtontimes.com/news/2014/jan/29/sochi -chief-city-is-worlds-most-secure-venue/print/ (accessed 19 April 2015).

Zhemukhov S (2010) Skating on Thin Ice in Sochi. *Open Democracy,* 4 June. Available at: www.opendemocracy.net/od-russia/sufian-zhemukhov/russia-skating -on-thin-ice-in-sochi (accessed 3 November 2014).

Zhemukhov S and Orrtung R (2014) Sochi Is the Scene of a Terrible Crime against Circassian People. *The Conversation,* 12 February. Available at: theconversation.com/sochi-is-the-scene-of-a-terrible-crime-agai nst-circassian-people-22913 (accessed 25 April 2014).

Part IV
Places and Sites

11

Where Conflict and Peace Take Place: Memorialization, Sacralization and Post-Conflict Space

Laura Michael, Brendan Murtagh and Linda Price

Introduction

This chapter is concerned with the disruptive potential of memory in peacebuilding processes where and when they materialize in the built environment. There is a diverse literature on how museums, memorials and sculpture are used to signify, electively narrate or even erase history and condensed forms of heritage (Schramm 2011). However, there is comparatively less work on how such processes confront mainstream policy communities concerned with place-making in uncertain and vulnerable post-conflict conditions. This analysis aims to evaluate the confrontation between public policy (in planning, urban management and development) with places that are loaded with meaning and memory for ethnic groups determined to legitimate their past as well as their future claims. It sets the context by conceptualizing the technical routines of planning and its concern with mediating interests, communicative action and collaborative practice, with the need to understand how space is socially constructed and, in particular, how memorialization elevates place from the mundane to the sacred.

This chapter argues that communicative practices have potential but, reduced to technorational processes, struggle to accommodate the complex and contradictory subjectivities involved in remaking the past in the service of the present and the future. Empirically, it focuses on the Maze Long Kesh (MLK) Prison in Northern Ireland, which was gifted to the newly established Northern Ireland Assembly after the 1998 Peace

Agreement brought an end to nearly three decades of violence. The MLK was built to hold both Loyalist (Ulster Volunteer Force and Ulster Defence Association) and Republican (Irish Republican Army (IRA) and Irish National Liberation Army (INLA)) paramilitaries, and it was where ten IRA and INLA hunger strikers died in 1981 in a protest over their status as political rather than criminal prisoners.

The Northern Ireland conflict centred on competing identities between a Protestant/Unionist/Loyalist bloc that broadly wants to maintain the union with Great Britain, and a Catholic/Nationalist /Republican community that mainly aspires to the reunification of Ireland. The 1998 Peace Agreement established new political institutions, equality legislation and mechanisms to release paramilitary prisoners committed of political offences. The new Northern Ireland Assembly is led by an executive, which involves the main political parties in a statutory coalition but is now dominated by the Democratic Unionist Party (DUP) and Sinn Féin, which was politically linked to the IRA. The executive is chaired by a first minister from the DUP and a deputy first minister from Sinn Féin. The development of the prison was taken forward by the Strategic Investment Board (SIB), established to oversee post-conflict reconstruction and to use any assets to attract inward investment, rebalance the economy (towards the private sector) and manage the delivery of a number of former military sites handed over to the local administration. The plan for the MLK proposed a mixed-use development but controversially retained parts of the prison as an interpretative centre along with a new building dedicated to peace, to be funded by the EU PEACE III Programme. For Republicans it is a 'sacred site of martyrdom' (Graham and McDowell 2007: 361), but for Unionists it is a 'shrine to terrorism' (*The Newsletter* 20 July 2012: 1), and ultimately these ethnic discourses could not be overcome by the power of capital, attempts at collaborative planning or a multitude of governance arrangements designed to bring the protagonists together.

The official response to the planning and development of the MLK drew heavily on collaborative planning methods, especially a concern for wide-ranging consultations with stakeholders, partnership approaches to key themes on prisoners, security interests and education groups, and an emphasis on exchanging ideas and experiences (Healey 2010). The methodology underpinning the case studies aligns with these concepts by mapping and interviewing the key actors, evaluating the performance of governance arenas, and assessing the quality of discursive argumentation and how it enabled the project to progress.

Ultimately, this all largely failed to resolve the contested nature of the site but it does not mean that discursive practices, governance and mediation are meaningless in such environments. Rather, this chapter highlights the need to reform such discourse around realist agonistic, as opposed to antagonistic, social relations, worked through governance regimes that have authority and some form of accountability. The MLK case demonstrates how memorialization competes with modernization in places coming out of conflict in increasingly unstable and disruptive ways. The chapter concludes by highlighting the normative value of agonistic strategies and in particular the importance of verifiable knowledge in creating and maintaining more authentic discursive practices.

Memorialization and reconstruction

Peacebuilding strategies and the struggle to maintain uncertain post-conflict stability have become increasingly concerned with the normalization of markets, reconstructing infrastructure and derisking regions for inward investment and growth (Richmond 2011). These liberal forms of peace fuse political agreement with opportunities for accumulation as donor strategies aim to 'transact' investment funds as a reward for restoring specifically capitalist behaviours and methods of working (Richmond and Mitchell 2012). Economic and physical planning are critical in such reconstruction processes, and rational ideologies inform local policy-making to ensure that the necessary infrastructure is put in place to enable competition and growth (Richmond 2011). There is no doubt that the immediacy of peacebuilding means that such discipline and organizational technologies are critical to the effective and efficient provision of services, goods and facilities that people genuinely need. The problem is that such technocratic regimes reify place in a way that can be redesigned through a set of rational procedures, rules and regulatory devices. A dominant and domineering economic discourse emerges, bolstered by professional planning systems and legal frameworks centred on private property rights, development codes and comprehensive zoning. However, as Yiftachel and Gahem (2004) argue, such certainties are at best limited and at worst harmful where space cannot simply be planned and designed in abstract and reductive ways. This is especially the case where conflict is territorialized and space becomes both a material and a non-material resource to be claimed, fought over, won and lost (Till 2012). Even when hostilities have ended and some form of accommodative politics emerges, space remains a potent reminder of what conflict was for and how central it remains to enabling or disabling post-conflict transition (Neill 2006).

In this respect, heritage becomes a political resource to bolster identity and a connection to a cause, especially where it can be concretized in material artefacts, museums and sites in what Till (2012: 3) calls 'wounded cities'. Place is thus inextricably linked to memory, memorializing and identity; it is the 'most serviceable reminder of what has happened in our past, what is happening right now and what may come in the future' (Lowenthal 1979: 110). Sites and artefacts thus help to reconstitute an 'imagined community' that binds people and place in the service of contemporary politics and ideological hegemonies, and in the marginalization of a disagreeable 'other' (Anderson 1983).

As such, places are open to an infinite number of interpretations, becoming 'a palimpsest of overlapping multi-vocal landscapes' that can readily slide into contested and contradictory narratives about history, specific events, victimhood and victory (Saunders 2001: 37). Monuments may no longer represent the 'right memory' but can become a collective social symbol with the ability to encapsulate and perpetuate ethnonational identities and claims (Schramm 2011). The temporal aspects of heritage sites are thus critical as time enables a shared experience to be mobilized in which people are connected to events, both physically and emotionally. Post-memory is as critical, by handing down through the generations a version of events that reproduces identity and anchors memory in the site of the most intense experiences (Greenspan 2005). However, Graham and Howard (2008) argue that such site-fixated heritage risks fetishizing place and obscures a wider social memory capable of accommodating different recollections and interpretations, a broader timescape, and the importance of intangible social markers and events. The MLK has little intrinsic or aesthetic heritage value as a fairly unspectacular 1970s brick-built prison but, as Graham and Howard argued, the tangible can only be interpreted through the intangible: 'Material heritage sites may comprise no more than empty shells of dubious authenticity but derive their importance from the ideas and values that are projected through them' (Graham and Howard 2008: 3).

These ideas and values can be dissonant and can reject official historical narratives to present an alternative reading of place and memory. The problem with 'dissonance arises because of the zero-sum characteristics of heritages, all of which belong to someone and logically, therefore, not to someone else. The creation of any heritage actively or potentially disinherits or excludes those who do not subscribe to, or are embraced within, the terms of meaning attending that heritage' (Graham and

Howard 2008: 3). This also risks a particularistic approach that limits the potency of a site as political heritage, especially outside dissolute interests and values. As Dwyer (2000: 667) argued in his consideration of the American Civil Rights Movement, the challenge is to 'jump scales' in order to make connections with other struggles and sites of oppression to externally validate memorial processes and the sanctity of a particular place or event. Such legitimization tactics are thus critical in addressing selective and even sectarian modes of remembering that are often associated with dissonant artefacts, museums and memorials.

The intensification of such memory-making is best achieved by elevating places to the sacred, and while pilgrimage trails and religious sites have emerged as a distinct heritage sector, the enshrinement of place is also tactically vital in asserting dominant ethnonational identities (Schramm 2011). Sturken (2004) argued that such sites develop a quasireligious, mystic status, becoming frozen, saturated with meaning and unable to exist in an everyday sense. The sacred is also manufactured as 'a cultural technique for creating sanctity... as a specific form of dealing with historical events' (Eschebach 2011: 134). The MLK is being socially and materially produced, and the interaction between memorialization and sacralization versus the technical specifics of redevelopment and planning reveal the disruptive capacity of place in post-conflict conditions. Planning theorists argue that such disruption needs to be anticipated, identified and built into decision-making processes. Difference is a sociospatial reality, and methods and processes that transform dissonance into productive forms of democracy are at the heart of collaborative planning (Innes and Booher 2010).

Collaborative planning

Allmendinger (2009) points out that the development of collaborative planning in the 1980s in the UK was a response to a frustration with technical rationality and the onslaught on the profession during the Thatcherite period. He also shows that it has moved planning beyond a preoccupation with land uses to become more concerned with the quality of discourse, governance and stakeholder relations in shaping spatial outcomes. Communicative theorists argue that different modes of reasoning and systems of meaning have equivalent status in debate, and the task for planners is to create a collective approach based on interaction and dialogue (Healey 1997). Here, language is vital, and in planning the priority is to establish a process of interactive collective reasoning or discourse which, in turn, involves a degree of collaboration, trust and reciprocity: 'In the end, what we take to be true and right will lie in the power of the better argument articulated in specific socio-cultural contexts' (Healey 1997: 54). These contexts value everyday experiences and an institutional approach, which recognizes that human actions and discourses are played out within the context of broader economic, labour-market and political structures. It suggests that individuals are not passive receptors of these systems but reflexive agents with the choices and capacity to modify and even transform the structuring forces that influence their lives (Innes and Booher 2010). Moreover, such practice can work across even the most divided contests and conflicts by acknowledging diversity, interdependence and dialogue within and between protagonists:

> Interdependence among the participants is the source of energy as it brings agents together and holds them in this system. Authentic dialogue is the genetic code, providing structure within which agents can process their diversity and interdependence… Diversity is the hallmark of the informational age. The wide range of life experiences, interests, values, knowledge and resources in society is a challenge for planning and the efforts to produce agreements and collective action.
>
> (Booher and Innes 2002: 227)

Critics point out that collaborative planning is theoretically limited because it focuses on a critical commentary about planning rather than a societal critique of planning (Gunder 2010). The emphasis should not concentrate on the conduct of planners and their practices but rather on the broader power structures and 'legitimisation dynamics within which public agencies often act' (Yiftachel 2001: 253). In short, because it does not question the authority embedded in power relations, it lacks the capacity for transformative change, especially for those at the fringes of decision-making. Allmendinger (2009) also points out that interests have different access to information, and can mobilize and interpret knowledge in vastly differing ways: between martyrdom and terrorism; sacralization and profanity; and history and a contrived version of heritage.

Knowledge and how it is used, corrupted and controlled is thus critical to authentic discourse, how claims are made and falsified, and how public officials understand their professional roles in managing conflict (Murtagh and Ellis 2011). The task is to acknowledge and domesticate 'antagonism' (irreconcilable conflicts and interests) into 'agonism', recognizing the inevitable competition in land uses that will ultimately impact upon the quality of life for the 'other' (Hillier 2002). Creating a competitive space in which interests bargain for recognition, precedence and acclaim does not mean that difficulties will disappear, but it will provoke crucial new ways of thinking about the nature of 'strife' in power relations. Strife is 'the expressive form of agonism' and places an emphasis on discourse and how meanings and interpretations can affect planning systems (Pløger 2004: 75). Pløger advocates participatory processes that stress openness, temporality, respect for difference, and the need to live with inconsistencies and contingency as a way to progress. Ways of achieving this in practice include open-ended processes, politically autonomous yet responsible institutional design, a plurality of discourses, and an ongoing mutual and critical dialogue between politicians, planners and citizens (Pløger 2004: 87). To deliver this, Flyvbjerg (2004: 295) argues that planning processes and methods should 'focus on values, get close to reality, emphasise little things', as well as studying cases and their contexts, including their sociopolitical meaning and how place is variously used as an economic, cultural and territorial resource.

Healey proposed a methodology for implementing and evaluating such an agenda in specific policy-making and planning environments. Her institutional audit unpacks the circumstances, settings and routines which might constitute agonistic, strife-driven relations into progressive local politics, bargaining and agreement. Here we need to work through a set of interconnected questions:

- Who has a stake in the qualities of the urban regions and how far are these stakeholders actively represented in current governance arrangements?
- In what arenas does discussion currently take place? Who gets access to these? Do they interrelate issues from the point of view of everyday life and the business world or do they compartmentalize them for the convenience of policy suppliers?
- Through what routines and in what styles does discussion take place? Do they make room for diverse ways of knowing and ways of valuing representation among stakeholders or do dominant styles dominate?
- Through what policy discourses are problems identified, claims for policy attention prioritized, and information and new ideas filtered? Do these recognize the diversity among stakeholders?
- How is agreement reached, how are such agreements expressed in terms of commitments and how is agreement monitored? Is it easy for those who are critical to implementation of the agreement to escape from the commitments? (Healey 1996: 213–214; 2010)

These can be used in an evaluative or normative sense and seek to offer a diagnostic tool both to understand and to rehabilitate conflict in land-use processes. By looking at each component we can understand where things go wrong as well as when and how they work. It is not a template but a heuristic that is used here to evaluate the attempts to develop and deliver a comprehensive and agreed plan for the MLK.

The MLK

Widely considered as one of the most significant material represen-tations of Northern Ireland's troubled history (Neill 2006), the MLK is located ten miles southwest of Belfast and housed approximately 25,000 prisoners and 15,000 prison staff over the period of the con-flict (Coiste na n-larchimi 2003). The 360-acre site was used until the 1960s as a Royal Air Force (RAF) base, and between 1971 and 1972 for the mass internment (without trial) of mainly Nationalists as a hastily conceived and counterproductive response to the outbreak of violence in the late 1960s (Flynn 2011). Internees were held in a series of old RAF compounds, but, as the conflict escalated, these were no longer suited to modern security needs. In 1976 a new prison complex was built based on a set of eight self-contained H-shaped blocks, which themselves became a symbol of British oppression and colonialist pun-ishment among Nationalists and Republicans (see Figure 11.1). MLK prisoners were effectively treated as prisoners of war, receiving special privileges on visits, freedom of association and the right to refuse prison uniforms. In 1976, however, this special category status was rescinded, leading to a series of protests both inside and outside the prison. The refusal to wear prison uniforms and dress only in blankets ultimately led to the decision by Republicans to commence a hunger strike, a tac-tic used throughout Irish history to mobilize popular, and especially international, opposition to British rule.

Methodology

Methodologically this case study draws on participant observation at four meetings of the reference groups set up to oversee the project; documentary analysis of political party statements, media coverage and written policies and programmes; and 18 semistructured in-depth inter-views with the key actors in the development process. These included managerial board members (see the Programme Delivery Unit below, n = 4 interviews); the planning and policy community (3); museum and heritage experts (4); users present on the wider site (2); political and

Figure 11.1 MLK Prison photographed shortly after its closure (Photo Brendan Murtagh)

prisoner interests (4); and the local community group (1). The institutional audit provides the framework to help conduct the observations, interrogate policy documents and design the interview schedule, but also makes explicit the positionality of the researchers and the way in which the data are collected and interrogated. Hermeneutic interpretations of sociohistoric events, especially in Northern Ireland, privilege the speaker but limit the capacity for challenge, dialectic entanglements and falsification. Researchers clearly need to gather the cultural knowledge to interpret and validate the experience of others and make explicit the frameworks that guide their methods and practice (Milner 2007). By adopting the institutional audit we clarify the variables and analytical categories that are critical to the formation, implementation and later collapse of the MLK project.

The context

The hunger strike in 1981 was phased, with the first IRA member, Bobby Sands, dying on 5 May. He was followed by nine others, each one associated with violent street protests in Republican areas, growing (critical) global interest and the politicization of a resurgent Sinn Féin. A total of 100,000 people attended Sand's funeral, by which time he had been elected as a member of Parliament to the British House of Commons, further ramping up pressure on the Thatcher administration. The hunger strike was called off in October 1981 and while the specific demands of the prisoners (such as the right not to wear prison clothes) were eventually granted, there was still no restoration or official recognition of their special category status.

In 1994 the IRA declared its first ceasefire but it was not until Good Friday in 1998 that a political agreement was finally reached between the political parties, including Sinn Féin. A central part of the agreement also involved the early release, on licence, of paramilitary prisoners, and in 2000 the prison finally closed. Ownership of the site was subsequently transferred from the British government to the Northern Ireland administration through the Office of the First Minister and Deputy First Minister (OFMDFM) in 2002. A number of former prisons, military bases and police stations were granted to the OFMDFM as part of the Reinvestment and Reform Initiative. This was designed to modernize infrastructure, regenerate decommissioned sites and strengthen private (preferably international) investment led by the newly formed SIB, which itself was modelled on the Kosovo Trust Agency (Flynn 2011). Within this context the overall management and control of the project was put in the hands of a new group called the MLK Programme Delivery Unit (PDU), and while it had no legal status it was part of the SIB and answerable to junior ministers in the OFMDFM.

A masterplan was developed proposing a sports stadium, event space for agricultural shows, an aviation museum based on the RAF's heritage and, crucially, and an international centre for conflict transformation. There was some uncertainty about what the centre would contain, how it would remember the hunger strike, who would decide and how rival interests would be accommodated in telling their different stories of prison life (Flynn 2011). Flynn also showed that it became quickly embroiled in disagreements about the location of the sports stadium and its access to Belfast, as well as its siting in an increasingly disputed and troubled project, not aided by the financial crises in 2008.

The masterplan was subsequently scaled back but £18 million of EU PEACE III Programme funding was secured for what was now termed the Peace-building and Conflict Resolution Centre (PbCRC). The shift in language to dedicate the centre to peacebuilding aimed to reframe the site as a place of learning and global research rather than to remember the specifics of the Republican struggle. A number of former prison buildings were retained and officially 'listed' as protected structures, including the prison hospital where the hunger strikers died, the main administration block, watchtowers, a church and one of the eight H-blocks. The remaining buildings were demolished, leaving the site available for sale for development as part of the wider regeneration scheme.

This section uses Healey's institutional audit to unpack the plan-making process, the positions of the stakeholders, methods of engagement, and the discursive strategies used to pursue respective interests and claims. Certainly the PDU did attempt to deliver a wide and inclusive process by clearly mapping the range of stakeholders from local communities, prisoners and guards, political parties, professional archivists and potential users. Separate arenas were established to accommodate each set of interests, canvass positions, develop proposals and explore options. While this enabled a broad debate about the site and its use, it also brought to the surface some deep sensitivities about the prison, its 'wounded' character, and the multiple and contradictory routines about hurt, time and morality. This was most evident in a discursive tension between official narratives that stressed the economic value of the site and its regeneration potential, and a sectarianized meta-narrative, which ultimately debilitated the process. Thus agreement was not reached and the site (like the various atrocities, parades and flags) 'fixed' the past in a way that was not reducible to this version of collaborative working. However, the analysis concludes by highlighting the value of the framework, not to reach consensus but to point to some of the critical components that might offer a more engaged approach to the use of knowledge, authenticity and validation.

The stakeholders

Figure 11.2 is important in this respect in mapping out the stakeholders in the use, management and development of the MLK. It shows the importance of political interests and especially the distinction between Republicans and Unionist as summarized by a representative from the academic sector: 'For Sinn Féin this is Robben Island, a misty-eyed look back on the heroic struggle... but for Unionist it is a shrine to terrorism and the attitudes are visceral. There is no way a deal could be done.' The statutory interests cut across planning, environmental management and heritage, and local residents groups were involved, as were victims

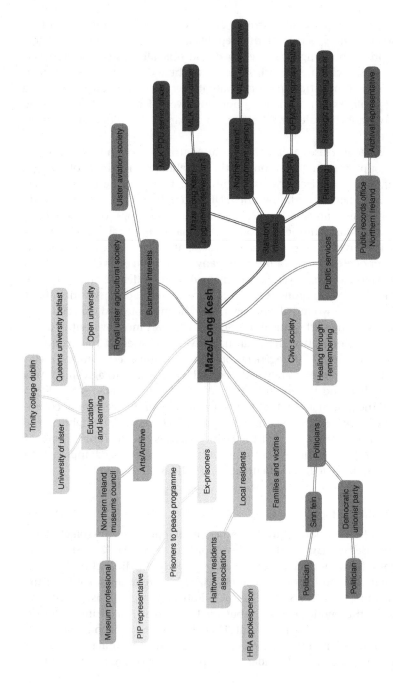

Figure 11.2 Collaborative stakeholder map for the MLK Prison

groups and prisoners' families. Prisoners and various ex-prisoner support groups, museum and tourism experts, as well as prison officers and security force representatives were also identified in the clusters. Finally, there was a set of business stakeholders with an interest in the wider development, and the universities that had delivered programmes in the prison and who were also viewed by the PDU as potential tenants of the peacebuilding centre.

Arenas for discussion

Six reference groups were established by the PDU and were organized around thematic interests: arts and archives; education and learning; civic society; public services; ex-prisoners; and victims and survivors. These groups consisted of around 15–20 members and did structure a comparatively open dialogue around technical issues, sensitive subjects, and how the development would accommodate the memories of the people most affected by the prison and its legacy.

> The Groups worked at one level... everyone was respectful and we got a chance to give our views. I heard things I hadn't thought of and it made me think about my position on it.
>
> (Ex-prisoner representative, Belfast, 21 February 2015)

However, such consultative structures were precisely that and, as will be shown later, they had little or no decision-making powers, access to resources, or control over the delivery and monitoring of the wider masterplan. It 'did move us to first base and consolidate the debate – we all knew what we were dealing with' (Academic representative, Belfast, 9 January 2014), and it allowed the risks to be understood in a way that enabled the development to move to the delivery stage. The working groups met regularly over a two-year period to define the issues, consider plans, advise the PDU on the way forward and, ultimately, endorse the overall masterplan. Their work was informed by various technical studies that included a number of business plans; financial analysis and investment appraisals; architectural surveys; design concepts; and a heritage review of how the retained buildings would present the prison's history. While there were issues about the scheme among loyalists and some right-wing Unionist politicians, there were no significant political objections and it was thus safe to move the project to the implementation phase.

For that to happen, a new structure that was smaller, more business-led, politically appointed and implementation focused was needed. The

MLK Development Commission (MLKDC) was established in October 2012 and very much modelled on the trusted semiprivate urban development corporations, which had led the regeneration of waterfronts across the USA and the UK since the 1970s. Announcing plans for the new vision and development plan (MLKDC 2013), the chairman stated that the scheme would generate £300 million of investment and create 5,000 new jobs:

> Our vision, *From Peace to Prosperity*, is intended to demonstrate how economic development can help consolidate and build upon our peace. We will promote the PbCRC as a showcase to attract international developers and investors. From the interest already shown, I am confident that we will have in place £100m investment by 2016.
> (MLKDC 2013: 1)

From Peace to Prosperity (MLKDC 2013) emphasized the commercial opportunities on the site, its proximity to the main Belfast–Dublin motorway and the anchor tenants, including the aviation museum and the agricultural society that were already committed to the project. Brochures, videos and commercial agents were employed to market the scheme that now carefully repositioned the PbCRC as an asset, able to connect to international researchers and resources. The use of the remaining H-block or hospital and how they might be integrated into the centre was not described or discussed in any detail in the vision and development plan document.

Routines

The shift from the PDU to the MLKDC was reflected in changes in decision-making styles and routines, and the elevation of commercial and economic imperatives that would presumably override mundane sectarian concerns. Aesthetics also became an important feature of such reframing and rescaling. The MLKDC commissioned Daniel Libeskind, a world-renowned architect, best known for his design of the Ground Zero memorial in New York, as the principal architect for the PbCRC. Libeskind is something of an expert in the architecture of trauma and had stamped his distinctive postmodern style on the Holocaust at the Jewish Museum in Berlin, as well as preparing the plan for the reconstruction of the World Trade Center. Ponzini and Nastasi (2011) highlight the value of 'starchitects' in conferring legitimacy on sponsors, creating a spectacle above the ordinary and revalorizing place, not for its functionality but for its symbolic value. However, Libeskind had to defend the scheme against accusations that it mimicked some of his earlier projects, and that he failed to reflect local sensitivities and narratives:

> I did not say that the design of the Centre should reflect any particular group's story. I listened extensively to the multiple perspectives presented and what I said was, 'all stories should be told', which has been everyone's goal from the start.
>
> (quoted in *Belfast Telegraph*, 10 May 2013: 1)

The education and learning working group focused, in particular, on the PbCRC as a potential international university research facility that could link the experiences of Northern Ireland with other sites of conflict and transformation. The centre aimed to draw in international scholars and researchers, further decoupling it from the raw intimacies of May 1981, but even here there were concerns about how different stories would be narrated. First, the PDU and then the MLKDC insisted that the term 'museum' was inappropriate and stressed 'peace' as the meta-narrative for the building. However, it was difficult to avoid the reality that the centre, and specifically the prison buildings, were primarily artefacts through which the political history of the Northern Ireland conflict would inevitably be interpreted. The heritage and tourism consultants brought in to advise on the complex suggested that a minimalist approach should be adopted to permit visitors to construct their own meaning and interpretation (and presumably avoid having to select what should be included and what should not be shown).

Discourses of memory

At the outset the principal stakeholders did acknowledge the sensitivities of the site and how to articulate its significance as an icon of the Troubles, but more importantly as an expression of a new time and a new politics. 'The site, to me is a microcosm of the peace processes' (Member, PDU, Belfast, 24 February 2014) and 'it's probably the biggest symbol of the Troubles' (Archival representative, Belfast, 29 January 2014). Attempts to articulate these very different stories, one hopefully compensating for the other in order to create the balance needed to progress the scheme, were not necessarily inconsistent. However, the semiotic significance of the H-blocks and how they are depicted on the wall murals of Belfast, memorials (including in the USA and France) and non-tangible culture, such as songs, poems and oral histories, is hard to resist: 'everyone's image of the site, is an aerial view of the site as an H block... that's what everybody thinks of the site' (Senior member, PDU, Belfast, 5 March 2014).

However, as much as it galvanized Republican politics, it also otherized and even marginalized different readings of both the Troubles and the prison. A significant discourse is shaped around the language of pain, with the MLK variously described as a place of 'raw', 'fresh', 'hurt'. Interviewees likened the events around the prison to a wound in social memory, reinforcing its sacred qualities with a PDU member explaining that 'there's rawness about this site' to a prisoner representative who stated that 'there's an awful lot of hurt there', referring especially to the families of prisoners (Prisoner representative, Belfast, 24 February 2014). There were a number of prison breakouts by Republicans, including the escape of 38 prisoners in 1983, when a prison officer died and six others were injured. Over the prison's 30-year history, a number of guards and security force members working there were killed by the IRA. Republicans did not oppose the inclusion of these deaths or the experiences of the state in the H-blocks, but the legacy of the prison in the lives of people who never actually went there demonstrates its potency, even in post-conflict Northern Ireland.

A distinctive discourse thus emerged around 'time', and specifically a concern that the rehabilitation of the site was too hasty, as one person explained: 'we are not ready for this yet' (Unionist politician, Belfast, 5 December 2013). The best time to tell the prison's history, if at all or under what conditions, was unclear, but officials also recognized that 'the passage of time is no bad thing perhaps' (Museums representative, Belfast, 6 March 2014) and 'it takes time, to go through that whole process, but it's a necessary time' (OFMDFM representative). Resonating with Moshenska's (2009) critique of the impacts of bombing and destruction upon memory, participants evoked wound-like imagery, stating that 'it is only closed a year, thirty years since the Hunger Strike, twenty years since people got out under the Good Friday Agreement, it's still raw for people' (Member, PDU, Belfast, 24 February 2014). Eschebach (2011) argued that living memory must expire in war sites before the question of how to tell their story can be dispassionately weighed, as summed by a community sector stakeholder: 'I think it's maybe just too, too fresh just to keep something... but time will tell' (Belfast, 8 November 2013).

Follow through to implementation

Ultimately, the quality and efficiency of collaborative planning processes are evaluated against what they produced as material consensual outcomes. Deyle and Wiedenman (2014) argued that the problem is that the start-up phase, with a wide identification of stakeholders, inclusive governance and multimethod consultation processes, involves comparatively easy commitments for an authority to make. Turning these initial debates into agreement required more manipulative and less collaborative tactics: 'you have to nurse it almost like an infant, and build up the trust... it was more of a black art than a science' (Member, PDU, Belfast, 24 February 2014). Moreover, the authority of the reference groups and their role in decision-making was also questionable, or at least unclear: 'They weren't actually making the decisions and we've been very clear that you know we refer to them and take their views into account' (Senior member, PDU, Belfast, 5 March 2014). Stakeholders were a 'resource' through which interpretations of the site were managed, but for some participants they descended into a mere 'talking shop' (Residents association representative, Halftown, 21 January 2014). As the process progressed, their status as stakeholders rather than passive consultees with various levels of interest in the project gradually unravelled: 'I think the level we were involved at wasn't at decision maker level, it was, "this is what we propose here, are you alright with that?" It wasn't involving, it wasn't a built up process' (Future site occupier, Belfast, 21 November 2013) and 'we don't have the same effect on the people who are holding the purse strings' (Site occupier, Belfast, 21 November 2013). The use of consultative processes to incorporate disagreeable interests is not new but ultimately the failure to deliver an agreed outcome rested, in part, on the quality of the engagement and seriousness of the discussion in the planning process.

Others were more cynical and saw the collaborative design purely in legitimation terms, and that 'there was a blueprint sitting in the cupboard that hadn't been brought out yet... there's always another agenda' (Residents association representative, Halftown, Maze, 21 January 2014). In short, there was a parallel process with the outward-facing collaborative design presenting an ordered, objective, difficult but ultimately inclusive attempt to produce agreement running alongside a hidden, politicized deal making a trade-off to secure the one thing that mattered – the Republican story of the hunger strike:

> I'm aware that it's almost like a dual management, they're managing the interest groups, and then obviously they take the feedback... I'm not too sure exactly what it's achieving.
>
> (Archival representative, Belfast, 29 January 2014)

Republicans said remarkably little publicly about the MLK, always down-playing the prison's history and elevating the peacebuilding centre and the value of the Northern Ireland process for states and societies coming out of conflict. The Loyalist Orange Order along with other Unionist politicians, including some from the DUP, formed a Raze the Maze campaign and gathered over 10,000 signatures on an online petition calling the demolition of the whole site. As the 2014 European and local government elections loomed, the DUP was under increasing pressure from Loyalists, other Unionist political parties and the Orange Order about its tacit endorsement of the project and formally withdrew its support in August 2013. This effectively ended the project, and shortly after the EU also withdrew the £18 million from the PEACE III Programme. The DUP leader published an open letter, which highlighted the fictitious nature of the opposition but maintained a veto over the development and especially the content of the retained buildings:

> Since this present deceitful Maze campaign began we have consistently and frequently stressed that we will not permit any shrine to be erected at the Maze and that no decision had been taken about the content and programme for any new Peace Centre or the use of the retained buildings, but that has not stopped our political opponents from inventing stories and seeking to frighten and raise concerns by agitating those who have suffered most from violent terrorism.
>
> (quoted in *Belfast Telegraph*, 15 August 2013: 1)

By 2015 the agricultural society had moved on to the site and the aviation museum is in the process of being developed, although hampered by a lack of funding. Sinn Féin and Republican ex-prisoners maintain that the PbCRC and the retained buildings will be developed, and Libeskind recently claimed that it would be simply a matter of time as all such toxic projects go through disagreement and negotiation before they are finalized.

Conclusion

A fresh set of interparty talks were established in late 2014 to deal with the past and parades, and the use of cultural symbols and the future of the MLK are now part of these negotiations. Ultimately its development will be sorted out by the ethnic poker that characterizes these arrangements, not collaborative planning regimes and discursive practices. Such wicked problems are not necessarily reducible to communicative deliberation, but the analysis does not render these approaches redundant, especially as a method in mediation and peacebuilding. The PDU and subsequently the MLKDC invested in collaborative concepts, but these were limited. Initially the authorities sought to foreground its economic value at the expense of its political history and contested meanings. The modalities of (neo)liberal peace underpinned the project by appealing to investors, highlighting its strategic location and branding it as an icon of modernization that had seamlessly displaced its violent past. There was also an attempt to decouple the project from the specifics of the site and re-present its meaning by jumping scales and appealing to its international, aesthetic and research significance.

Graham and McDowell (2007) are wrong to dismiss the MLK as the product of a manipulative and selective Republican statecraft. These places, and there are others, cannot be erased or devalued because they do not comply with the reimaging of Northern Ireland as a liberal, economically progressive and socially readjusted place. But nor is it helpful to see them endlessly reproduce a form of sectarian ethnonationalism; elements of the process have implications for the management of places and events with such disruptive potential in peace processes. Some stakeholders are more important than others, but the process did not reveal or debate (in a discursive sense) what the explicit positions of ex-prisoners, families, prison officers, security forces and victims were on the actual development. Their claims about the MLK needed to be more openly presented, and the basis of such claims made transparent and open to verification and falsification. As the discourse descended into predictable sectarianism, the claims about the prison and the peacebuilding centre became more hysterical, unfounded and visceral. The DUP's first minister admitted as much in the open letter that effectively collapsed the project. The centrality of knowledge, its validity and reliability, and the capacity of others to unpack it would not necessarily have saved the development but it would have made clearer the nature of claims and counter-claims and how to evaluate their relative merits. The working groups were supported with largely technical and especially financial information, but lay knowledge and the lived experiences of prisoners and victims does not mean that such evidence inevitably becomes inferior or sectarian.

Handling such sensitivities requires a different range of skills and competencies among planners and project managers, especially when they work within state structures, cultures and laws. Professional routines may have leant on the collaborative, but how to identify the conditions of strife and how various forms of power are constitutive of agonistic relations requires a stronger engagement with the methods of peacebuilding and conflict transformation. Such interdisciplinarity will not guarantee success either but it does highlight the need to rethink how a range of professionals, especially those involved in place-making, are central to political stability. In particular, there is a need to see such processes, methodologies and skills move outside community relations, political mediation and the peace sector, and into the mainstream cultures, practices and educational frameworks of the range of professionals charged with spatial development.

References

Allmendinger P (2009) *Planning Theory.* Basingstoke: Palgrave Macmillan.

Anderson B (1983) *Imagined Communities.* London: Verso.

Belfast Telegraph (2013) *Design Row over Daniel Libeskind's Maze Peace-Building Centre in Northern Ireland.* Belfast: *Belfast Telegraph*, 10 May.

Belfast Telegraph (2013) *Maze U-Turn: Peter Robinson's Letter in Full to Senior Party Figures.* Belfast: *Belfast Telegraph*, 15 August.

Booher DE and Innes JE (2002) Network Power in Collaborative Planning. *Journal of Planning Education and Research* 21(3): 221–236.

Coiste na n-larchimi (2003) *A Museum at Long Kesh or the Maze?* Belfast: Coiste na n-larchimi.

Deyle RE and Wiedenman RE (2014) Collaborative Planning by Metropolitan Planning Organizations: A Text of Casual Theory. *Journal of Planning Education and Research* 34(3): 257–275.

Dwyer OJ (2000) Interpreting the Civil Rights Movement: Place, Memory, and Conflict. *The Professional Geographer* 52(4): 660–671.

Eschebach I (2011) Soil, Ashes, Commemoration: Processes of Sacralization at the Former Ravensbrück Concentration Camp. *History and Memory* 23(1): 131–156.

Flynn MK (2011) Decision-Making and Contested Heritage in Northern Ireland: The Former Maze Prison/Long Kesh. *Irish Political Studies* 26(3): 383–401.

Flyvbjerg B (2004) Phronetic Planning Research: Theoretical and Methodological Reflections. *Planning Theory and Practice* 5(3): 283–306.

Graham B and Howard P (2008) Introduction: Heritage and Identity. In Graham B and Howard P (eds) *The Ashgate Research Companion to Heritage and Identity.* Aldershot: Ashgate, pp. 1–18.

Graham B and McDowell S (2007) Meaning in the Maze: The Heritage of Long Kesh. *Cultural Geographies* 14(3): 343–368.

Greenspan E (2005) A Global Site of Heritage? Constructing Spaces of Memory at the World Trade Center Site. *International Journal of Heritage Studies* 11(5): 371–384.

Gunder M (2010) Planning as the Ideology of (Neoliberal) Space. *Planning Theory* 9(4): 298–314.

Healey P (1996) Consensus-building across Difficult Divisions: New Approaches to Collaborative Strategy Making. *Planning Practice and Research* 11(2): 207–216.

Healey P (2010) *Making Better Places: The Planning Project in the Twenty-First Century.* Basingstoke: Palgrave Macmillan.

Hillier J (2002) Direct Action and Agonism in Democratic Planning Practice. In: Allmendinger P and Tewdwr-Jones M (eds) *Planning Futures: New Directions for Planning Theory.* London: Routledge, pp. 110–135.

Innes JE and Booher DE (2010) *Planning with Complexity: An Introduction to Collaborative Rationality for Public Policy.* Abingdon: Routledge.

Lowenthal D (1979) Age and Artefact: Dilemmas of Appreciation. In: Meinig D (ed.) *The Interpretation of Ordinary Landscapes: Geographical Essays.* New York: Oxford University Press, pp. 103–128.

MLKDC (2013) *From Peace to Prosperity Vision and Development Plan.* Belfast: MLKDC.

Milner HR (2007) Race, Culture, and Researcher Positionality: Working through Dangers Seen, Unseen and Unforeseen. *Educational Researcher* 36(7): 388–400.

Moshenska G (2009) Resonant Materiality and Violent Remembering: Archaeology, Memory and Bombing. *International Journal of Heritage Studies* 15(1): 44–56.

Murtagh B and Ellis G (2011) Skills, Conflict and Spatial Planning in Northern Ireland. *Planning Theory and Practice* 12(3): 349–365.

Neill WJV (2006) Return to Titanic and Lost in the Maze: The Search for Representation of 'Post-conflict' Belfast. *Space and Polity* 10(2): 109–120.

Pløger J (2004) Strife: Urban Planning and Agonism. *Planning Theory* 3(1): 71–92.

Ponzini D and Nastasi M (2011) *Starchitecture Scenes, Actors and Spectacles in Contemporary Cities.* Turin: Umberto Allemandi.

Richmond OP (2011) Critical Agency, Resistance and a Post-colonial Civil Society. *Cooperation and Conflict* 46(4): 419–440.

Richmond O and Mitchell A (2012) Introduction. In: Richmond O and Mitchell A (eds) *Hybrid Forms of Peace: From Everyday Agency to Post-Liberalism.* Basingstoke: Palgrave Macmillan, pp. 1–38.

Saunders N (2001) Matter and Memory in the Landscapes of Conflict: The Western Front 1914–1919. In: Bender B and Winer M (eds) *Contested Landscapes: Movement, Exile and Place.* Oxford: Berg, pp. 37–53.

Schramm K (2011) Introduction: Landscapes of Violence: Memory and Sacred Space. *Heritage and Memory* 23(1): 5–22.

Sturken M (2004) The Aesthetics of Absence: Rebuilding Ground Zero. *American Ethnologist* 31(3): 311–325.

The Newsletter (2012) 20 July, *Maze Shrine Now a Step Closer.* Belfast: *The Newsletter.*

Till KE (2012) Wounded Cities: Memory-Work and a Place-Based Ethics of Care. *Political Geography* 31(1): 3–14.

Yiftachel O (2001) Can Theory Be Liberated from Professional Constraints? On Rationality and Explanatory Power in Flyvbjerg's. *Rationality and Power. International Planning Studies* 6(3): 251–255.

Yiftachel O and Ghanem A (2004) Understanding 'Ethnocratic' Regimes: The Politics of Seizing Contested Territories. *Political Geography* 23(6): 647–676.

12
Seeing and Unseeing the Dome of the Rock: Conflict, Memory and Belonging in Jerusalem

Nina Fischer

Introduction

The summer and autumn of 2014 were Jerusalem's most conflict- and violence-ridden period in recent years.[1] After the 1 July funeral of Naftali Fraenkel, Gilad Shaer and Eyal Yifrach, three teenage yeshiva students who were kidnapped and killed by Palestinian militants in the West Bank, groups of Jewish Jerusalemites took to the streets looking for revenge, shouting 'Death to Arabs' and attacking those they could find. Later that night, extremist Israelis abducted 16-year-old Mohammad Abu-Khdeir from outside his home in Shu'afat, a neighbourhood in East Jerusalem. The next day, it became known that he had been burned alive. As a reaction to the murder and the rise in overt racism against Arabs, violent protests erupted all over Jerusalem's Palestinian areas. The clashes with the police did not subside until October and violence claimed the life of 16-year-old Muhammad Sunuqrut from Wadi Joz,[2] while hundreds of East Jerusalemites, including minors, were arrested. But even when the large-scale protests decreased, attacks and deaths on both sides continued, as fear, anger, hatred and loss of hope became pervasive among all inhabitants of the city.

Many Jerusalemites and others who love the city have shared their knowledge and thoughts with me. Especially important were Dareen Ammouri, Hagai Barnea, Osama Elewat, Daniel Feldman, Ronny Ishaky, Yossi Malca, Wendy Pullan, Ariela Ross-Jayosi, Natasha Roth, Kate Solomon and Mirella Yandoli. A special thank you goes to the women of the Ammouri family for introducing me to the Haram al-Sharif as a site of living Palestinian culture.

Among the triggers bringing Arab youths onto the street are settlers taking over houses in Palestinian neighbourhoods and the announcements of new Israeli settlements in the eastern parts of Jerusalem. These settlements are illegal under international law, although not Israeli law, and occupy land that Palestinians hope will form the capital of their state. The most explosive spark, however, is any act seen as an attack against the compound in the Old City, where the al-Aqsa Mosque and the Dome of the Rock stand, and which Gershom Gorenberg poignantly calls 'the most contested piece of real estate on earth' (2000: 11).

The site is known in Arabic as Haram al-Sharif – the Noble Sanctuary – and colloquially as the Haram or the al-Aqsa compound; while in Hebrew, it is called Har HaBeit – the Temple Mount.[3] Muslims believe that it is from this site that the Prophet Mohammad took his night journey to heaven, while in the Judeo-Christian tradition, this marks the location of the former Jewish Temples. The compound's status as a sacred site, which plays a role in the memory of the three monotheistic religions, makes it a particularly complicated problem in any Israeli-Palestinian peace negotiation. Indeed, Ron Hassner argues throughout his work on Jerusalem that the indivisibility of sacred spaces further entrenches the conflict's intractability (Hassner 2003, 2009).

The al-Aqsa compound is within the boundaries of Jerusalem, and Israel controls security and access, but it is administered by the Jerusalem Islamic Waqf. This charitable trust stayed under Jordanian control even after 1967, meaning that the site is run neither by the Israelis nor by the Palestinians. This arrangement, which gives the Hashemite Kingdom a voice in Jerusalem practically and politically, introduces another dimension into the conflict, while at the same time reminding us that the Haram and its shrines are religious symbols for Muslims around the world.

Palestinians in particular are fearful of any change to the *status quo* of this sacred site – the third-holiest site in Islam – because the general belief of the Palestinian street is that the ultimate Israeli goal is to destroy the Dome of the Rock and the al-Aqsa Mosque.[4] This is not least due to the machinations of extremist Jewish groups such as the Temple Mount Faithful, whose explicit aim is to build the Third Temple. In the 1980s, such aspirations led to a plot by the Jewish Underground, a terrorist group inside the settler movement, to blow up the Dome of the Rock, which was ultimately stopped by the Israeli security services (Gorenberg 2000: 128–137). Exacerbating such fears, throughout 2014, more and more Jewish visitors, including prominent Members of the Israeli Knesset (MKs) entered the grounds of the complex.[5] The problem is not the visitors as such but that many are on the far right of the political spectrum and hold extremist ideas concerning the rebuilding of the Temple. Israeli security forces increased restrictions on Palestinian access[6] and violent clashes erupted on the compound, with videos of tear gas being shot into the mosque making their rounds on the Internet, enraging viewers. As a result, on 17 October the Palestinian Authority President, Mahmoud Abbas, called on Palestinians to unite and protect the Haram: 'it's not enough to say that the settlers have come; we must stop them from entering the compound by any means necessary' (Ma'an 2014a). This statement, for Palestinians, is tantamount to a call to defend Jerusalem. Israel interpreted Abbas' comments as incitement to violence, while MKs such as Miri Regev added fuel to the fire by publicly promoting a Knesset bill to change the *status quo* on the site to allow non-Muslim prayer.[7]

The Dome of the Rock, belonging and denial of belonging in Jerusalem

In this chapter I will suggest that in conceptions of the Haram al-Sharif/Temple Mount religion, collective memory, national identities, notions of belonging and (urban) politics conflate, as is evident in the use of visual symbols by both Palestinians and Israelis. Jerusalem is an intrinsic part of, and utilized in, the political conflict, due to its role in collective memories and identities of Israelis and Palestinians, and their respective sense of belonging in this city. I maintain that the Dome of the Rock is a central icon of Palestinian identity and a visual symbol of the Arab past and present in Jerusalem, also for Israelis.

The Dome of the Rock, in its visual uses in particular, also refers to the Haram as a whole, which, along with the al-Aqsa Mosque, act as a 'call to action' symbol for Palestinians and Muslims worldwide. According to the Quran, the Haram is 'the farthest mosque' – al-Aqsa – and the location of Mohammad's night journey. The evidence is still visible: the stone which the Dome of the Rock encircles is believed to hold the footprint of Mohammad.[8] The site also appears in Muslim collective memory as the first *qibbla* (direction of prayer) during the prophet's life. But the Haram is a symbol of resistance for Christian Palestinians as well. The numbers of Christians may be small overall in the West Bank and Gaza but they form a significant part of the Jerusalem ID card-holder community and if Muslim Palestinians were looking for a purely religious symbol, the much better choice would be the al-Aqsa

Mosque, which, unlike the Dome of the Rock, is mentioned in the Quran.

This conflation of memory, belonging and politics, as well as the conflict it evokes, were actualized for me in one situation to which I was repeatedly drawn during the year 2014 which I spent in Jerusalem Muslims praying on the street outside Damascus Gate and the Old City. They worshipped next to vegetable stalls, buses and trash cans, at the feet of Israeli forces facing them from behind barriers in full riot gear, with the golden cupola of the Dome of the Rock peeking out from behind the famous walls of Jerusalem. What stuck in my mind in particular was the first Friday of the 'holy month' of Ramadan on 4 July, when access to the al-Aqsa Mosque was age restricted; Israel expected clashes because Mohammad Abu-Khdeir's funeral was planned for after noon prayers. On first impression such scenes might seem to be only about religion. After all, these are worshippers. However, since it took place in Jerusalem – the fault line of the national struggle – far more is implied.

Everyone involved – the praying men, the journalists and bystanders like myself, and the security forces – knew that this prayer and all the others being undertaken in Jerusalem's streets were intent on highlighting the fact that access to the al-Aqsa compound is denied for these Jerusalemite worshippers and thus such prayers are also non-violent acts of resistance against Israel and its policies. At the same time, worshipping in front of barriers and highly armed forces is a ritualized performance demonstrating belonging in a contested space (Butler 1990). The prayers in this location also emphasize the significance of the Haram as a venerated Palestinian space that embodies symbolic meaning, both religious and national.

Any Palestinian act of non-violent resistance is encapsulated in the concept of *sumud*, meaning steadfastness or perseverance. The primary icon of *sumud* is the olive tree because of the rootedness it symbolizes, but it has a double meaning: it is both an ideology of physically staying on the land, drawn from a deep sense of belonging to it – individually and collectively – , and a political strategy of non-violently resisting the occupation. Raja Shehadeh, a Palestinian lawyer and writer, explores the concept as a form of cultural resistance in his memoir *The Third Way* (1982) as an intentional way of life in which every act is informed by a refusal to accept the *status quo* of occupation and to lose more land.

The notion of belonging connected with questions of space and place has drawn responses across the disciplines, including anthropology and human geography (Frykman 1999: 14). For Palestinians, the Haram al-Sharif, visualized in the Dome of the Rock, more than anything symbolizes their city and their existence in it. Here, belonging is literally space-bound in a long history that includes centuries of Arabs continuously calling Jerusalem home, despite past and current incidents of displacement and dispossession, a conflict-ridden present and an unclear, worry-inducing future. Edward Relph argues that 'to be inside a place is to belong to it and to identify with it, and the more profoundly inside you are the stronger is this identity with the place' (1976: 49). *Sumud* is the Palestinian identification with and belonging to a place from the inside – it is lived and it is performed to show resistance.

Such a demonstration of belonging also plays itself out in pictorial representations of the city. The Dome of the Rock is respectively called on by Palestinians or suppressed in some Israeli representations in order to establish and project origins in and ownership of Jerusalem, by highlighting one's own or delegitimizing the other side's narrative of the city; in this case, a spatial narrative of famous Islamic architecture identifying Jerusalem as a city with an Arab history and present. Palestinians, in particular Jerusalemite Palestinians, use the image of the golden dome ubiquitously in a visual performance of belonging to this place. Similar to the Damascus Gate prayer, place itself is at the centre of this performance of belonging to Jerusalem, for both internal and external consumption.

On the Israeli side, the Dome of the Rock, symbolizing the non-Jewish element of Jerusalem, is increasingly edited out of municipal and other official representations of the city. I argue that the authorities are thus giving an implicit answer to Palestinian national aspirations and claims to Jerusalem: their place in the city is visually denied. Simultaneously, some Palestinians have publicly questioned the former existence of the Temples on the compound, casting doubt on a Jewish past in Jerusalem and implicitly denying Israeli/Jewish claims to the city.

The Israeli denial of the Arab presence is not unlike the Palestinian historical revision. However, Palestinians are not in power, and we therefore encounter a top-down versus bottom-up movement in the use of images that are more than just visual entities: they represent the imbalance of conflict in Jerusalem. In spatial theory, 'the right to the city', introduced by Henri Lefebvre, is still an amorphous concept used to advocate rights for all by social movements, for the anti-capitalist struggle, or to describe the right to suitable urban spaces for all inhabitants (Purcell 2002: 101). I want to suggest that the politicized use of visual symbols of cities, how they manifest belonging and who is allowed access needs to be part of this debate.

Pictorial representations of the city as microcosms of conflict

My corpus includes pictorial representations of Jerusalem used – by individuals and collectively – on posters, graffiti, city signs and souvenirs, to name but a few. While thinking about contested space, I draw on memory and visual culture theory, mainly to theorize my questions concerning using urban images as a political act and a performance of belonging. In the context of this collection on spaces of conflict and peacebuilding, what we currently observe in Jerusalem is a petrification and extension of conflict in the visual symbols employed. The research on conflict and Jerusalem is, of course, wide and multidisciplinary, yet cultural studies approaches that pay attention to the spatial and visual are rare. To the best of my knowledge, only Dana Hercbergs and Chaim Noy study images of the city as symbols (2013).

WJT Mitchell suggests that since 9/11, a virtual 'war of images' (2011: 2) has erupted, as iconic images – be it of the 'Falling Man' of the Twin Towers or the 'Hooded Man' of Abu Ghraib – have become part of cultural archives and shape our perceptions of the world. Dora Apel refers to a 'contest of images' in our contemporary culture of war, which inserts images of suffering into the local and global media (2012). Also, in the case of Jerusalem, different visual representations of the conflict – be it the praying men on the street, the suffering caused by state power or the effects of terror attacks – are inserted into this war. Thus both Israelis and Palestinians hope to have their respective victim status acknowledged.

Similarly to Mitchell and Apel, I position my chapter within a war of images – although a different version of it, since I am looking at much quieter images, which are not splattered with blood but show urban sites to make a claim about ownership and belonging as a spatial narrative of Jerusalem's past and present is drawn up. An image of the Dome of the Rock does not induce the same sense of helplessness that images of human suffering cause (Sontag 2003: 100) but rather works on a different level; it is a cultural icon symbolizing the religious, national and ethnic belonging of Palestinians in Jerusalem.

Mike Parker has defined certain characteristics for cultural icons: they are images that are distinct, durable and reproducible, and that 'reside in the collective memory of large groups of people' (2012: 12) which form a 'receptive' community for 'discernible tragic-dramatic narratives' revealed by the cultural icon (2012: 13). I take recognizable buildings and other architectural elements as cultural icons that tell a story of their city, referencing layers of urban history, and that are key to understanding spatial narratives of memory that foster a sense belonging.

Parker's definition brings together the visual and memory studies aspects of my methodology because, effectively, I also understand the Haram (visualized in the Dome of the Rock) as a site of memory – that is, an element of the past that is used to give meaning in the present, creating a focal point of collective identity. Or, in the words of Pierre Nora, such a site is 'any significant entity, whether material or non-material in nature, which by dint of human will or the work of time has become a symbolic element of the memorial heritage of any community' (1996: xvii). In this specific case, we encounter a site of memory that is a site in the literal sense of the word, since here, memory, its collective meaning and its visual representation come together in an actual, lived-in space that denotes belonging in the sense of Relph (1976).

Collective forms of memory bind communities together, as a common past is the basis for ethnic or national groupings, much as we see in the meanings ascribed to the Temple Mount/Haram al-Sharif. Memory always tends towards spatialization, as Edward Casey asserts in his concept of 'place memory':

> It is the stabilizing persistence of place as a container of experiences that contribute so powerfully to its intrinsic memorability. An alert and alive memory connects spontaneously with place, finding in it features that favour and parallel its own activities. We might say that memory is naturally place-oriented or at least place-supported.
>
> (1987: 186–187)

Memory cannot be disregarded in the conflict between Israel and Palestine, particularly not in Jerusalem, where it is politicized on a daily basis in a struggle over the place that the respective side's memory narratives are bound to. Maurice Halbwachs demonstrated how memory of place, in particular of Jerusalem, is rewritten according to a certain group's narratives (1941). His study of the Christian topography in the Holy Land presents it as socially constructed. Such constructions and uses of narratives of the space of Jerusalem are continually changing, often done intentionally to suit the political purposes of both Palestinians and Israelis.

David Freedman has shown that images are powerful, and not only when depicting atrocities. Indeed, he finds that they elicit a range of emotions, among them 'admiration, awe, terror and desire' (1989: 433). The war of images over the Dome of the Rock as a site of memory and a cultural icon with a spatial history of Arab Jerusalem elicits sentiments of belonging to, and ownership of, space. This is not just a competing representation of city images; the image of the golden cupola represents the physical space and the human presence of the Palestinian population of Jerusalem. Showing the Dome of the Rock is for Palestinians a form of *sumud*, of place-taking in the city, and Israeli images that block it out implicitly deny an Arab place in, and right to, the city.

Seeing the Dome of the Rock

When one walks through the Old City quarters inhabited by Arabs (Muslim and Christian), the streets of East Jerusalem, and elsewhere in Palestine, images of the Dome of the Rock are everywhere: on shop windows, posters and even pictures mounted on houses.[9] The golden cupola decorates general commerce (e.g. printed on plastic shopping bags, tissue holders, containers of produce); it appears as the logo of the Bank of Palestine and many shops; and it is used on products aimed at a more global audience of tourists via souvenirs (e.g. embroidered bags, postcards and key chains). The ubiquity of the image holds a persistent message of belonging to, and a historical connection with, the Haram al-Sharif. Caliph Abd al-Malik built the Dome of the Rock in 691 and, apart from the 88-year Crusader rule (1099–1187) when the building was turned into a church until Saladin's takeover, it, the Haram and the city were always in Arab hands.

The emphasis on the Arab presence in Jerusalem is also evident in other artistic representations of the golden dome, as expressed through Palestinian art, film and literary genres from life-writing to fiction and poetry. In memoirs we learn most about the Haram's role in Palestinian culture. Exemplary is *In Search of Fatima: A Palestinian Story* by Ghada Karmi, whose family fled Jerusalem in 1947. When Karmi sees TV footage of a victorious Israel in 1967,

> Memories, dormant for years, of visiting my aunt's house in the Old City and playing with other children in the giant forecourt of the mosque on hot, still afternoons stirred inside me. The vast tiled courtyard in front of the Dome of the Rock used to make a perfect playground for hopscotch, and the historic arches, pillars and holy sanctuaries were ideal for games of hide-and-seek. As there were no parks or open spaces in the Old City where children could play, the mosque compound was the only place available.

(2002: 370)

Karmi highlights a double loss of home city and a way of life, losses which ultimately caused her political engagement. Thus the Haram is portrayed as a cultural/national icon and a lived-in social space in the sense of Lefebvre (1991), rather than solely a sacred site.

William Safran presents defining factors for expatriate communities, such as ethnicity, minority status and a consciousness of peoplehood. Among those he sees 'a continuing orientation to a homeland and to a narrative and ethnosymbols related to it' (2005: 39). Sites of memory play a role in collective memory and nation-building processes and, if they are spatial ethnosymbols, they strengthen the connection between the group and its place of origin.

The need for strong narratives and symbols to ensure ethnic cohesion is particularly true for Palestinians; as a community with many different fractures (those living in the diaspora, the internally displaced, those living as Israeli citizens, East Jerusalemites and those in the West Bank and Gaza), they are, in the words of Benedict Anderson, an 'imagined community' in its most literal sense (1983). Not only that, but in public discourses and political negotiations, the notion of a Palestinian homeland is contested and frequently rejected, in particular when it comes to the status of Jerusalem, the issue that has derailed all peace talks so far. Edward Relph argues: 'a deep human need exists for association with significant places' (1976: 147). His considerations concerning 'placelessness' resonate with the situation of the Palestinian people, the majority of whom are not in what they consider their historic homeland, and yet their sense of belonging has not diminished. In the case of the Dome of the Rock, which tells a spatial narrative of belonging, the 'association with a significant place' is striking because here, Palestinians living in Jerusalem, and those far from it, place themselves within the city by displaying the iconic image. While not able to walk the streets of the city, *sumud* – in a visual performance of belonging – is possible even from a distance, which is in line with Anne-Marie Fortier's considerations regarding diaspora identities between belonging and longing to belong (2000). But given the unresolved political situation, the Palestinian performance of belonging in Jerusalem always has the added aspect of non-violent resistance.

Representations of the golden cupola also appear in explicitly political ways, with every official pronouncement of Palestinian Authority, politicians and speakers shot against the backdrop of Jerusalem, with the Dome of Rock centred in the middle. Here, of course, the narrative runs along the lines of: 'Look at this, it is a formidable building of Islamic architecture, it shows our long-term presence in this city, which we think of as our capital.' Palestinians attribute the saying 'without Jerusalem there is no Palestine and without Palestine there is no Jerusalem' to Yasser Arafat.

The image does not appear in the logos of the older political parties, such as Fatah (1959/1965), the Palestine Liberation Organization (1964) and the Popular Front for the Liberation of Palestine (1967), but all of them were established before Israel took control of the Old City. However, the logos of military factions such as the al-Aqsa Martyrs' Brigade, Islamic Jihad in Palestine and Ezzedin al-Qassam Brigades, which were all founded later, all carry the icon. This implies that the compound was originally constructed as less national and more religious. Today, the secular parties also make use of the Dome of the Rock, thus reinforcing it as a national symbol.

Another example of this national adoption is that every image of a 'martyr' – whether it is a Palestinian victim or someone who committed an attack – is decorated with it. Even the huge poster hanging on the wall of the Abu-Khdeir family home in East Jerusalem's Shu'afat, which commemorated their lynched son, had the face of Mohammad inset into an image of the golden dome.

Another location for the golden cupola is the Israeli West Bank barrier. In Jerusalem and other urban areas, it manifests as 6 to 8 meters high concrete slabs, carrying urban art and political slogans, with the most recurrent image depicting the Dome of the Rock. Indeed, the wall might cut Palestinians off from Jerusalem and access to the Haram al-Sharif, but the sheer number of images gives evidence that longing, sense of belonging and resistance – here in cultural form – are not diminished by concrete obstructions.

Robert Sauders maintains that the graffiti on the wall 'transformed what was previously considered an everyday act of nationalist, non-violent resistance and internal political communication to a complex performance of international solidarity and diverse global social justice communication' (2011: 16). The pictorial language of the golden cupola translates to an international audience, and the image is indubitably used as a way to call on international solidarity. For me, as an outsider and yet someone who has walked the streets of East Jerusalem and along the wall uncountable times, alongside Palestinians, Israelis, and internationals, the Dome of the Rock – due to its beauty and visual ubiquity – always has a special impact. But these graffiti hold more: especially in the Jerusalemite neighbourhoods that are cut off from other areas of the city by the wall, they offer a reassurance of belonging and *sumud*, even as the conflict is a lived part of city life.

Similarly, most houses on the eastern side of the wall, at least those in the Jerusalem Governorate of the Palestinian Authority, carry a carved image of the Dome of the Rock in the stone over the front door. It is the symbol of the connection to Jerusalem, even if physical access is denied – we need to remember that only 5 per cent of Palestinians living in the West Bank have permits to enter Israel.

These are just a few examples of the appearance of the Dome of the Rock within Palestinian life. Many others could be cited, such as social networks on popular culture. However, my point concerns the visual pervasiveness of this cultural icon. Ultimately, this raises the question: Why do Palestinians need to insist on and remind themselves and others of their presence in Jerusalem and their right to the city, which is 'like a cry and a demand' (Lefebvre 1996: 158)?

Nira Yuval-Davis maintains: 'belonging tends to be naturalized, and becomes articulated and politicized when it is threatened in some way' (2006: 197). The repetition of the spatial narrative of the Arab past and present in Jerusalem, demonstrated in the Dome of the Rock, it is a articulation of belonging or, to bring in Edward Relph again, is a demonstration to being an insider identifying with the contested space (1976). Memory scholars have argued that repetition, such as ritual and re-enactment, strengthens memory in its collective formations (cf. Connerton 1989). But here, memory is also a form of resistance. It is *sumud* in its most space-bound meaning of performing belonging in Jerusalem, which is triggered not only by a fear of erasure but by history: Palestinian collective memory is built around the experience of territorial losses to Israel during the *Nakba*, the Six-Day War and later dispossessions. In Jerusalem, 1948 meant losses in the mixed neighbourhoods in West Jerusalem; in 1967, annexation of the rest of the city, petrified in the Israeli Basic Law in 1980 that officially declared all of Jerusalem as the unified and indivisible capital of the State of Israel.

Fear and mistrust of Israel's intentions are part and parcel of Palestinian identities as their right to the city, in its most literal way, is always put into question. This is codified in the Zionist trope 'A land without people for a people without a land', which, in the words of Edward Said, is an example of Israeli hopes to 'cancel and transcend an actual reality – a group of resident Arabs – by means of a future wish – that the land be empty for development by a more deserving power' (1979: 9).[10] Furthermore, Palestinian losses are continuing as Israeli policy keeps Jerusalemite Palestinians in a tenuous status: residency rights are revoked if someone spends too long outside the city; the wall divides neighbourhoods; settler activity is increasing while building permits are hard to come by;[11] and house demolitions are on the increase.[12] The Palestinian 'right to the city' is literally at stake, both for individuals and collectively, and Israel has yet to agree to a capital of the future State of Palestine within the 'eternally indivisible' Jerusalem.

Unsurprisingly, Palestinian fears around the Haram al-Sharif/Temple Mount – given the respective religious claims – concerning dispossession or even destruction are particularly pronounced. However, for tension to climb, plans do not have to be as spectacular as the 1980s plots to blow up the Muslim shrines mentioned above; attempts to change the *status quo* are enough. Some right-wing Israeli activists frame their push for Jewish religious activity on the compound as a form of coexistence, but Palestinian experience in Hebron's Ibrahimi Mosque/Cave of the Patriarchs has shown that sharing a sacred space means dividing it, as the mosque in which settler Baruch Goldstein killed 29 worshippers in 1994 was divided between Muslims and Jews following the attack. Michael Dumper has warned of a 'Hebronization' of Jerusalem, 'the process by which the protection and development of Jewish holy sites also becomes the vehicle or bridgehead for further encroachments on Muslim property in Jerusalem by the Israeli state' (2014: 139). At the same time, Palestinians tend to conflate the intentions of extremists, the government and average Israelis, assuming that they all aim for the destruction of the Dome of the Rock and the al-Aqsa Mosque – a feeling presumably not quelled by the absence of these buildings in official representations of the city.

However, the focus on the golden dome and on the Palestinian spatial narrative of the Haram al-Sharif, which is only too understandable from the perspective of the occupied, also entails a tunnel view that sees only the city's Arab past and present. While it is not surprising that Palestinians do not depict Israeli cultural icons, such as the Knesset, some circles also question the existence of the Jewish Temples before the Haram became the Haram. For instance, Yasser Arafat argued at Camp David in 2000 that the Temple stood in Nablus, not Jerusalem (Ross 2004: 694). Similar attitudes are common in Palestinian public discourse – for example, in the *Ma'an* newspaper, which tends to use phrases such as 'the alleged Jewish Temple' or 'where Jews believe their Temple once stood', when discussing the al-Aqsa compound.[13] This can be read as an unseeing of the Temple and a refusal to acknowledge Jewish memory and belonging, historical evidence, Islamic scripture and local tradition which never questioned the Temples' existence until the eruption of the struggle between Arabs and Jews.[14] Or, in the words of Wendy Pullan and Brit Baillie, 'when conflict becomes extreme, plurality, and especially diversity, is one of the primary qualities to be rejected' (2013: 3). As we will see, this statement holds just as true for the Israeli authorities' approach to Jerusalem and its cultural icons.

Unseeing the Dome of the Rock

Using the Dome of the Rock to visualize Jerusalem is not new in cultural history. Indeed, the cupola was an iconic image throughout the history of Western art, including Jewish representations of the city. Examples of this can be found in medieval maps of the world, a selection of which are shown on the website of the Jerusalem municipality. Among them are the works of Bruno von Breydenbach (1486) and of Rabbi Chaim Salomon Pinta of Zefat (1875).[15] This suggests that before the struggle over land and over the city, the fact that this magnificent building is a Muslim shrine was not significant. Today, however, if we look at official and economically driven representations of the city, there is an increasing movement towards the use of cultural icons in the Western part of the city.

Dana Hercbergs and Chaim Noy (2013) identified a new spatial and visual regime – a 'Davidization' of Jerusalem in images used by tourism and real-estate industries. This development is linked with the stagnation of the peace process and with what has been called the 'Judaization' of Jerusalem, primarily the 'Holy Basin' – the Old City and its surrounding areas. Hercbergs and Noy demonstrate an insistence on replacing what was previously a double focus on the Dome of the Rock/Western Wall with the Tower of David and other symbols related to King David who made the city the capital of his kingdom, such as the lyre mirrored in the Chords Bridge (2013).[16] This argument has much traction with my thinking, as I suggest a similar westward movement in the Israeli use of symbols. Their focus, however, is not on the significance of the Dome of the Rock and deals with Jewish representations of the city alone.

What I see in official representations of the city is, rather than understanding them solely as replacing symbols, an explicit editing out of the cultural icon of the Other and thus of Jerusalem's Arab past and present. In the following section, I will discuss several examples to show the range of ways in which this happens. Most prominently, this is evident in the new city logo presented in January 2012. It brings together the traditional city emblem, its motto and a colourful mosaic of architectural elements of an entirely Jewish Jerusalem, among them the Chords Bridge, the Hurva Synagogue Arch,[17] the Montefiore Windmill,[18] the YMCA tower[19] and the Tower of David.[20]

The official emblem of Jerusalem, designed by Eliyahu Koren and depicted next to the mosaic of abstract images of buildings, also uses Jewish cultural icons. Under the Hebrew name of the city is a shield (presumably David's) with the rampant Lion of Judah (the symbol of the tribe and the Kingdom of Judah of which the capital was Jerusalem – fusing the two Jewish rules over the city), framed by an olive branch symbolizing the quest for peace. The background could be the Western Wall and the Old City ramparts, both of which were under Jordanian rule and inaccessible to Jews in 1950 when the symbol was adopted. Nowadays the emblem comes with the catchy phrase 'It's great/worthwhile to live in Jerusalem', which in Hebrew reads '*shaveh likhiot be-yerushalayim*'. However, *shaveh* has two meanings: it can be both 'worthwhile/great' (as it is in this sentence) or 'equal', which in the context of this chapter on Jerusalem's conflict, where equality is always at stake, brings in a questionable double meaning.

This latest logo replaces one that was used for less than a year – also a city skyline of famous buildings but, unlike any previous logo, it included the Dome of the Rock. In the current one the dome has disappeared again, even though, according to a city representative, the reason behind this most recent logo showing the buildings in a rainbow of colours is to represent the 'human, cultural, and emotional' diversity of Jerusalem life through its 'monumental buildings' (Municipality of Jerusalem Spokesperson 2012), which seems more than a little ironic considering that we do not see Jerusalem's most monumental building. Jewish symbols are part of an Israeli tradition of Hebraizing the land. Amneh Badran argues that Israel has always used them 'as signs of Jewish sovereignty', and the first example she gives, before going on to national symbols, is the municipality's lion. She continues: 'in contrast, the history and symbols of the Palestinians were either undermined or denied' (2010: 41) much like what happens with the Dome of the Rock in the municipal images.

Versions of the logo appear in different urban contexts, as official placards welcoming visitors or advertising events. All of them show different versions of a skyline including buildings of West Jerusalem and no 'Arab' ones. These municipal posters are also in Hebrew, with some English mixed in, but no Arabic, Israel's second national language, even though partially funded by city taxes.

Similarly, the skyline used on Jerusalem's garbage trucks excludes the Dome of the Rock, even though the majority of workers in this industry are Palestinians. In this image the Lion of Judah has morphed into a smiling toy pet with a broom, standing in a field of flowers, against a backdrop of only Jewish/Israeli cultural icons of Jerusalem. The Hebrew slogan on the trash truck – also lacking any Arabic – says: 'We are going for a clean city.' If read in the given context of repressed or cleansed city imageries, this has a disturbing undertone.[21]

However, it is not just the Jerusalem municipality that has 'disappeared' the Dome of the Rock. In 2012 the left-wing newspaper *Ha'aretz* reported that the Israel Defense Forces (IDF) Military Rabbinate released an educational package for Hanukkah, where one document showed the Temple Mount with the Dome of the Rock and the Fakhriyya minaret erased (Cohen 2012).[22] The IDF Spokesperson's Office claimed that this was done to represent Jerusalem during the Second Temple period and lambasted the newspaper for implying anything else. However, why leave in the visually less prominent and less culturally iconic al-Aqsa Mosque if that is the case? Another example is a municipal kindergarten near Haifa presenting their students for Passover 2015 with the Haggadah from which the Dome of the Rock had been erased (Tessler 2015).

Anyone flying out of Israel is also treated to similar images by an Israeli Airports Authority poster. On every Ben Gurion gangway it sends travellers off with wishes for a peaceful departure set against postcard-worthy pictures of famous sites in front of an Israeli flag. These visuals souvenirs are Jerusalem, Tel Aviv, the Stalactite (Avshalom) Cave, Caesarea, Haifa, the Dead Sea, the Banias Waterfall in the occupied Golan Heights, and Eilat.

Jerusalem, the first and biggest image, depicts three different locations: the Knesset, the Montefiori Windmill and the Western Wall shot in such a way that the Dome of the Rock, which crowns the Temple Mount above it, is invisible. The picture is not digitally altered but it is challenging to find an angle from which only the wall is in the frame. Even if a picture is taken close to the wall, the minaret will always be visible, but its top, with the half-moon symbol of Islam, is cut off here. This poster, too, with its solely Jewish narrative, shows that according to official representations, this country has no Arab spaces in Jerusalem or elsewhere.

What we see in all these examples is an exercise in forgetting or, more specifically, a repression of a non-Jewish cultural icon, an Arab narrative of Jerusalem and of Palestinian space. It should be noted that this is not a general tendency of individual Israelis, as one can see in endless souvenirs that represent the Dome of the Rock or in criticisms such as Eldad Brin's attack against the Jerusalem skyline used to promote the Jerusalem Formula 1 (2014). At the same time, though, those longing for a rebuilt a Jewish Temple, or hating Arabs, will see no Dome of the

Rock.[23] The official propensity to exclude Palestinian Jerusalem from view is an interesting reversal of philosopher Renan's assertions about forgetting and nationhood: 'Forgetting, I would even go so far as to say historical error, is a crucial factor in the creation of a nation, which is why progress in historical studies often constitutes a danger for [the principle of] nationality' (1990: 11). But here, the forgetting encouraged is of another nation which lives just metres away and which asserts its belonging to Jerusalem in the city's most prominent building, a cultural icon excluded by those in power.

However, WJT Mitchell argues in *What Do Pictures Want?* (2005) that images cannot be destroyed and, if an attempt is made to obliterate, or as we see in this case, to repress, an image, it will survive anyway because of images' tendency towards remediation: due to the workings of collective forms of memory, images do not disappear. Visually repressing the golden cupola is certainly not a violent destruction as we have seen in Iraq (e.g. in the Saddam Hussein statue being pulled down), but it has a similar intention. And yet this is not how pictures work. Michael Taussig, going even further than Mitchell, shows that any attempt at 'destroying an image, brings that image to life' (1999: 147), meaning that it has the paradoxical result of enhancing said image's life, and in Jerusalem the dome-shaped absence speaks volumes.

Especially given Israeli policies and the realities of conflict in the city, it is hard not to read the official representations as validating Stuart Hall's conception of how power shapes seeing, at least for a public that is not Palestinian and engaged in constantly performing belonging by representing the Dome of the Rock. Indeed, it seems as if the municipality wants to show an image of a Jerusalem without an Arab past and present by not portraying this spatial narrative, cutting out those who should not be seen, implying that they do not belong. As the ruling entity, Israel has the 'power to mark, assign and classify, power to represent someone or something in a certain way' (Hall 1997: 259). The official images show that Jerusalem is conceived of as wholly Jewish.

As mentioned earlier, Freedman argues that images elicit emotions, and not seeing the Dome of the Rock can be interpreted as acknowledging the significance of the site to the national identity of Palestinians, which threatens Jerusalem's Jewishness. Ultimately there is a paradoxical double message to such representations. Showing the city as only Jewish visually contradicts the official claim of a 'unified' Jerusalem that is Israel's 'eternal and indivisible' capital if large sections of the city are missing because, after all, 37 per cent of Jerusalem's residents are Arab.[24] Moreover, the absence of Arab Jerusalem in the official city logo raises another significant question: Does the exclusion mean that

East Jerusalem is considered not part of the city and is consequently Palestinian?

Conclusion: Placing the peace in Jerusalem?

As I sit here, drafting the conclusion, I have been hearing helicopters roaming the sky for almost 24 hours, a more extreme situation than what Jerusalem has witnessed all through the summer. These helicopters are my personal indicator of whenever the security situation – primarily in the Palestinian areas – takes a turn for the worse, and last night, on 30 October 2014, the situation reached a climax when a young Jerusalemite Palestinian associated with Islamic Jihad shot Yehuda Glick, one of the staunchest activists of the Temple Mount Faithful. Subsequently, as a reaction, Israel closed all access to the Haram al-Sharif/Temple Mount. Only the Palestinian director and the security forces were allowed in, but otherwise, not even the muezzin of the al-Aqsa Mosque was granted access. Only hours after the closure, this led to the declaration of a general Palestinian strike, fighting in the Old City which is expected to spread throughout East Jerusalem, and Mahmoud Abbas announcing that 'Israel is declaring war against the Palestinian people and their holy places and against both the Arab and Islamic nations' (Ma'an 2014b). A comprehensive closure of the compound has not happened since 1967 and the shockwaves were felt everywhere in the city. While I worked on this chapter, Jerusalem was burning, sometimes with big flames, sometimes at a smoulder, but the future is bleak.

In their study *Memory Culture and the Contemporary City*, Uta Staiger and Henriette Steiner argue:

> memorial and heritage sites are often annexed in order to promote particular interpretations of the past, thus enacting symbolic claims on the urban environment. More often than not, this renders competing narratives invisible, eliding the often fractured memory culture yielded by a single place or city.
>
> (2009: 7)

The Haram al-Sharif/Temple Mount is a striking example of how a site of cultural memory comes to signify contested belonging and singular narratives: within the context of contemporary Jerusalem politics and lived realities, even just an image of the golden cupola has become a space in which conflict erupts, is prolonged and petrified.

Brent Plate has pointed out that to understand visual products, we need to pay attention to a number of aspects, among them the historical context in which an image was created and the viewer's identity (2002: 5). The historical context of today's Jerusalem is one of high tension, conflict, occupation and an increasing drive towards violence rather than peacebuilding, while both Israelis and Palestinians are tightly holding onto their respective memory frameworks as demonstrated in the political uses of spatial cultural icons to promote a singular narrative while erasing the Other. If we think within terms of a 'right to the city' in the case of Jerusalem, it is not metaphorical; it is about who has the right to be represented, to see and to be seen, which ultimately leads to the question of who has the right to be in this city, even if it is asked by way of cultural icons and not outright. Memory is the basis for a collective identity and sense of belonging, and in the case of the Dome of the Rock it takes the form of a spatial narrative, which is then performed in the use of visual products. Images showing or not showing iconic buildings of the Arab past and present demonstrate who belongs or is placeless here in the sense of Edward Relph (1976). We need to consider such indirect ways of performing ownership and belonging, especially in Jerusalem, where the 'right to the city' has real-life consequences for Palestinian residents: between 1967 and 2010, some 14,034 had their residency status revoked and thus literally lost the right to live in their city (UNOCHAOPT 2011).

We currently observe Israelis and Palestinians foregrounding their respective narratives. However, this contested land, and Jerusalem more than any other place, has a complex set of stories, and these are complicated by religious beliefs being turned into collective memory and political argument. Uri Avnery, a founder of the Israeli peace movement, calls on us to 'demolish the myths, conventional lies and historical falsehoods, on which most of the arguments of both Israeli and Palestinian propaganda rest. The truths of both sides are intertwined into one historical narrative that does justice to both. Without this common basis, peace is impossible' (2010: 40). Also, peacemaking requires the acknowledgment of the other side's differing narrative (Bar-Tal and Haverin 2013; Bar-Tal et al. 2014). Today, however, conflict and separation are entrenched beyond policy and bloodshed, even in visual representations of the city.

While my analysis has focused on Jerusalem's reality of conflict being sustained, by default it also shows that peace cannot take place unless we pay attention to the multiplicity of images, and thus the belonging and equality of all of Jerusalem's residents. Don Mitchell sees social action, such as the taking of public spaces, even by marginalized groups, as a chance for a 'more open, more just, more egalitarian society' (2003: 10). Acknowledging the cultural icons of both sides is a metaphorical variant of such an optimistic taking of public spaces. When all are seen, the war of images ends: all buildings, those who belong to them and to whom they belong have a 'right to the city'.

Notes

1. As I check the copyedits of this chapter in October 2015, a new wave of violence is engulfing Jerusalem. The current attacks, which mainly involve stabbings and the shooting of the (suspected) assailants, are once again said to have erupted because of Palestinian fears concerning Israeli intentions on the al-Aqsa compound.
2. On 31 August, Israeli forces shot Sunuqrut in the head with rubber-coated steel bullets, and on 7 September he succumbed to his injuries. The teenager is said to have left his home to pray and shop for his family but got caught between protesters and armed forces.
3. In order to acknowledge both narratives about the compound, I will use the terms interchangeably, depending on what group and memory framework I am speaking about.
4. A Palestinian Centre for Policy and Survey Research (PSR) opinion poll, released on 9 December 2014, found that 86 per cent of Palestinians think that the Haram is 'in grave danger'. Fifty-six per cent believe that Israel intends to destroy the Muslim shrines and build a Temple and 21 per cent believe that Israel will divide the compound and build a synagogue (2014).
5. The Chief Rabbinate of Israel and different ultra-Orthodox streams explicitly forbid Jews from ascending on the mount in case they inadvertently step on the Holy of Holies: the exact location of the Temple is unknown.
6. Data released by the Jerusalem police to the NGO Emek Shaveh revealed that in 2014, Muslim access to the Haram was restricted (primarily restricting the entry of men under 50) on 41 days (i.e. 15 per cent of the year) in contrast with 3 days in 2012 and 8 days in 2013 (2015).
7. It is important to note that MK Regev is a member of Benjamin Netanyahu's Likud Party and therefore part of the Israeli mainstream. She is not considered particularly right wing or extremist. In Israel's 34th government (formed in 2015) she is the minister of culture and sport.
8. Christians believe that these are Jesus' footprints. In the Judeo-Christian belief system, this stone is the 'Foundation Stone' of the world; it is believed to be the site of Abraham's binding of Isaac and where Jacob, while escaping Esau, slept and dreamed of wrestling with the angel; and it is also believed to be the Temple's centre stone.
9. The Dome of the Rock is also to be found in other locations of the Arab (Muslim) world, showing an identification process with the Palestinian cause, but the use of the image is possibly indicative of a more religious perception of the al-Aqsa compound.
10. The sentence, most famously used by Israel Zangwill and Golda Meir, has received much critical attention. Israeli historian Anita Shapira, for instance, has explored its role in Zionist discourses (1992, 41ff.). Edward Said and Rashid Khalidi (1997: 101) have both studied the implications for Palestinians.

11. The problem is multilayered. Only 13 per cent of East Jerusalem has been earmarked for Palestinian construction (while 35 per cent of East Jerusalem land has been expropriated for Israeli settlements), thus severely limiting the possibility of even applying for a permit – permits are only given for areas which are zoned for Palestinian buildings (UNOCHAOPT 2011). In addition, the organization Bimkom – Planners For Planning Rights found that between 2005 and 2009 only 55 per cent of building permits for the Palestinian neighbourhoods where approved. In the Israeli areas it was 85 per cent (Association of Civil Rights in Israel 2015).

12. House demolitions are enforced either for building without permit or as a punitive measure against those who have committed terror attacks, meaning that even if the perpetrator had died or was imprisoned, the family home would be destroyed. In 2014 some 98 buildings were demolished, leaving 208 Palestinian Jerusalemites homeless (Association for Civil Rights in Israel 2015).

13. www.maannews.com. For me, a much stronger questioning of the former existence of the Temples became blatant on a tour of the Western Wall Tunnels organized by the Centre of Jerusalem Studies of al-Quds University, where the guide implied throughout that the famed Jewish buildings never stood in this location.

14. The current Palestinian disregard for the Jewish veneration of the Temple Mount is also evident in archaeology and the excavations performed by the Waqf, in particular those between 1996 and 1999 in the Solomon's Stable area, for which bulldozers were used and which caused an international outcry. At the same time, Israel is using archaeology to strengthen its narrative of the city, in particular in the contentious City of David dig and the Western Wall Tunnels. This has led to an accusation by Palestinians and critics around the world claiming that foundations of the Haram and the Old City are being damaged. The struggle over excavations is continuous and deeply contentious.

15. https://www.jerusalem.muni.il/en/Jerusalem/CapitalofIsrael/AncientMaps/Pages/default.aspx (accessed 22 September 2014).

16. The Chords Bridge, designed by Santiago Calatrava, was opened 2008.

17. The Hurva Synagogue in the Jewish Quarter was built in the 18th century, then destroyed and rebuilt in 1864. After Israeli forces withdrew from the Old City in 1948, the Arab Legion blew up the synagogue. Plans for reconstruction were made from 1967, but initially only a commemorative arch was erected. The rebuilt synagogue was dedicated in 2010.

18. The landmark windmill was built in the first Jewish neighbourhood outside the Old City walls in 1857. British Jewish banker Moses Montefiore funded it and today it is a museum commemorating him.

19. Lord Pulmer, the British high commissioner, dedicated the Jerusalem International YMCA in 1928. It was designed by Arthur Loomis Harmon and opened in 1933.

20. The Tower of David is a citadel just inside the Old City walls by Jaffa Gate. Its current shape dates back to the Mamluk and Ottoman periods, though parts are much older. Its tower was originally the minaret of an Ottoman mosque installed in the complex, then used as a garrison. Today the building houses the Tower of David Museum of the History of Jerusalem.

21. I asked some locals about this image. Israelis whom I consider mainstream-left responded that the caption has disturbing implications but was surely not planned. All Palestinians and far-left Israelis interpreted it as being used intentionally by the authorities.
22. One of the four minarets of the al-Aqsa, built in 1278 in the southwest corner of the Haram.
23. Long-term Jerusalem resident and critic Tom Powers has pointed to this tendency, particularly in art pieces sold by (ultra)nationalist Israelis in the Jewish Quarter of the Old City, who thus visualize a reality in which a rebuilding of the Temple is imminent (2012).
24. Numbers from December 2012, published in May 2014 by the Israeli Central Bureau of Statistics (2014).

References

Anderson B (1983) *Imagined Communities: Reflections on the Origin and Spread of Nationalism.* London: Verso.

Apel D (2012) *War Culture and the Contest of Images.* New Brunswick: Rutgers University Press.

Association for Civil Rights in Israel (2015) *East Jerusalem 2015 – Facts and Figures.* Available at: https://www.acri.org.il/en/2015/05/12/ej2015/ (accessed 2 July 2015).

Avnery U (2010) *Truth against Truth: A Completely Different Look at the Palestinian-Israeli Conflict.* 3rd edn. Tel Aviv: Gush Shalom.

Badran A (2010) *Zionist Israel and Apartheid South Africa: Civil Society and Peace Building in Ethnic-National States.* Abingdon: Routledge.

Bar-Tal D and Halperin E (2013) The Nature of Socio-Psychological Barriers to Peaceful Conflict Resolution and Ways to Overcome Them. *Conflict and Communication Online* 12(2): 1–16.

Bar-Tal D et al. (2014) Sociopsychological Analysis of Conflict-Supporting Narratives: A General Framework. *Journal of Peace Research* 51(5): 662–675.

Brin E (2014) Formula 1 promo de-Arabizes Jerusalem skyline. Available at: http://972mag.com/formula-1-promo-de-arabizes-jerusalem-skyline/97402/ (accessed 7 October 2014).

Butler J (1990) *Gender Trouble: Feminism and the Subversion of Identity.* London, New York: Routledge.

Casey E (1987) *Remembering: A Phenomenological Study.* Bloomington: Indiana University Press.

Central Bureau of Statistics, State of Israel (2014) *Media Release: Selected Data on the Occasion of Jerusalem Day (2012–2013) Data.* 25 May 2014. Available at: http://www1.cbs.gov.il/www/hodaot2014n/11_14_134e.pdf (accessed 8 July 2015).

Cohen G (2012) IDF Rabbinate Edits out Dome of the Rock from Picture of Jerusalem's Temple Mount. *Haaretz.* 5 January 2012. Available at http://www.haaretz.com/news/national/idf-rabbinate-edits-out-dome-of-the-rock-from-picture-of-jerusalem-s-temple-mount-1.405602 (accessed 8 July, 2015).

Connerton P (1989) *How Societies Remember.* Cambridge: Cambridge University Press.

Dumper M (2014) *Jerusalem Unbound: Geography, History, and the Future of the Holy City*. New York: Columbia University Press.

Emek Shaveh (2015) *Denial of Access and Worship on the Temple Mount/Haram al-Sharif in 2012–2014*. Available at: http://alt-arch.org/en/wp-content/uploads/2015/06/Denial-of-Access-and-Worship-2012-Eng.pdf (accessed 8 July 2015).

Fortier AM (2000) *Migrant Belongings: Memory, Space, Identities*. Oxford: Berg.

Freedman D (1989) *The Power of Images: Studies in the History and Theory of Response*. Chicago: University of Chicago Press.

Frykman J (1999) Belonging in Europe: Modern Identities in Minds and Places. *Ethnologia Europaea* 29(3): 13–24.

Gorenberg G (2000) *The End of Days: Fundamentalism and the Struggle for the Temple Mount*. New York: Free Press.

Halbwachs M (1941) *La topographie légendaire des Evangiles en Terre sainte: Etude de mémoire collective*. Paris: Presses universitaires de France.

Hall S (1997) *Representation: Cultural Representations and Signifying Practices*. London: Open University.

Hassner R (2003) To Halve and to Hold – Conflicts over Sacred Space and the Problem of Indivisibility. *Security Studies* 12(4): 1–33.

Hassner R (2009) *War on Sacred Grounds*. Ithaca: Cornell University Press.

Hercbergs D and Noy C (2013) Beholding the Holy City: In Italics Changes in the Iconic Representation of Jerusalem in the 21st Century. *Quest: Issues in Contemporary Jewish History* 6: 237–263.

Karmi G (2002) *In Search of Fatima: A Palestinian Story*. London: Verso.

Khalidi R (1997) *Palestinian Identity: The Construction of Modern National Consciousness*. New York: Columbia University Press.

Lefebvre H (1991) *The Production of Space*. Translated by Nicholson-Smith D. Oxford: Blackwell.

Lefebvre H (1996) *Writings on Cities*. Translated by Kofman E and Lebas E. Oxford: Blackwell.

Ma'an News Agency (2014a) Abbas Urges Palestinians to Defend Aqsa. 14 October 2014. Available at: http://www.maannews.com/Content.aspx?id=733724 (accessed 8 July 2015).

Ma'an News Agency (2014b) Abbas: Closure of Aqsa 'Declaration of War.' 30 October 2014. Available at: http://www.maannews.net/eng/ViewDetails.aspx?ID=736338 (accessed 8 July 2015).

Mitchell D (2003) *The Right to the City: Social Justice and the Fight for Public Space*. New York: Guilford Press.

Mitchell WJT (2005) *What Do Pictures Want?: The Lives and Loves of Images*. Chicago: University of Chicago Press.

Mitchell WJT (2011) *Cloning Terror: The War of Images, 9/11 to the Present*. Chicago: University of Chicago Press.

Municipality of Jerusalem Spokesperson (2012) Press Release: 'Yerushalaim, kav pirsumei hadash' (Jerusalem's new logo). Available at: http://cityncountrybranding.com/2012/01/09/חדש-פרסומי-קו-ירושלים/ (accessed 12 August 2014)

Nora P (1996) Introduction to Volume 1: Conflicts and Divisions. In Nora P (ed.) *Realms of Memory: The Construction of the French Past. Volume 1: Conflicts and Divisions*. New York: Columbia University Press, pp. 21–26.

PSR (2014) Public Opinion Poll (54), released on 9 December 2014. Available at: http://www.pcpsr.org/en/node/505 (accessed 8 July 2015).

Parker M (2012) *Cultural Icons: A Case Study Analysis of Their Formation and Reception*. Unpublished PhD thesis, University of Central Lancashire, UK.

Plate SB (2002) *Religion, Art, and Visual Culture: A Cross-Cultural Reader*. New York: Palgrave Macmillan.

Powers T (2012) The Disappearing Dome of the Rock. *Tom Powers – View from Jerusalem*. Available at: http://israelpalestineguide.wordpress.com/2012/01/19/the-disappearing-dome-of-the-rock (accessed 8 July 2015).

Pullan W and Baillie B (2013). *Locating Urban Conflicts: Ethnicity, Nationalism and the Everyday*. Basingstoke: Palgrave Macmillan.

Purcell M (2002) Excavating Lefebvre: The Right to the City and Its Urban Politics of the Inhabitant. *GeoJournal* 58: 99–108.

Relph E (1976) *Place and Placelessness*. London: Pion.

Renan E (1990) What Is a Nation? In Bhabha HK (ed.) *Nation and Narration*. Abingdon: Routledge, pp. 8–23.

Ross D (2004) *The Missing Peace: The Inside Story of the Fight for Middle East Peace*. New York: Farrar, Straus and Giroux.

Safran W (2005) The Jewish Diaspora in a Comparative and Theoretical Perspective. *Israel Studies* 10(1): 36–60.

Said EW (1979) *The Question of Palestine*. New York: Times Books.

Sauders RR (2011) Whose Place Is This Anyway?: The Israeli Separation Barrier, International Activists and Graffiti. *Anthropology News* 52(3): 16.

Shapira A (1992) *Land and Power: The Zionist Resort to Force, 1881–1948*. Oxford: Oxford University Press.

Shehadeh R (1982) *The Third Way: A Journal of Life in the West Bank*. London: Quartet.

Sontag S (2003) *Regarding the Pain of Others*. New York: Farrar, Straus and Giroux.

Staiger U, Steiner H and Webber A (eds) (2009) *Memory Culture and the Contemporary City: Building Sites*. Basingstoke: Palgrave Macmillan.

Taussig M (1999) *Defacement: Public Secrecy and the Labor of the Negative*. Stanford: Stanford University Press.

Tessler I (2015) Dome of the Rock Erased from Kindergarten's Haggadah. *Ynet News*. 31 March 2015. Available at: http://www.ynetnews.com/articles/0,7340,L-4642864,00.html (accessed 8 July 2015).

UN Office for the Coordination of Humanitarian, Occupied Palestinian Territory, (UNOCHAOPT) (2011) *Key Humanitarian Concerns*. Available at: http://www.ochaopt.org/documents/ocha_opt_jerusalem_factsheet_december_2011_english.pdf (accessed 8 July 2015).

Yuval-Davis N (2006) Belonging and the Politics of Belonging. *Patterns of Prejudice* 40(3): 197–214.

13
Belfast, 'The Shared City'? Spatial Narratives of Conflict Transformation

Milena Komarova and Liam O'Dowd

Introduction

This chapter explores an emerging narrative of Belfast as a 'shared city' and its potential to contribute to the transformation of territorial divisions between ethnonational communities. While the 'shared city' idea ostensibly manifests the search for political accommodation and social reconciliation, we explore *de facto* conflicting understandings, enactments and spatialities of sharing the city and ask: What visions of 'peacebuilding' does this narrative expose? In what types of public space has sharing been possible? How is sharing performed in these spaces? And what new inclusions and exclusions are these performances producing?

Policy agenda and academic discourse in the UK over the past couple of decades (Holland et al. 2007; Home Office 2001) have often focused on the centrality of public space (particularly in cities) in fostering 'community cohesion', 'integration' and social interactions. Public agency and think-tank reports (Lownsbrough and Beunderman 2007) have also emphasized the significance of public space as a social resource in regeneration policies and in creating 'sustainable communities'. They have discussed the conditions for the workability of individual types of public space: how well these can be made to resonate with everyday life routines, and how successfully they accommodate wider definitions of 'community'. In many ways this discourse was spurred by the aftermath of the race disturbances in a number of British cities at the turn of the century and was a response to a perceived crisis of a shared public realm (Nagel and Staeheli 2008; Phillips et al. 2007). On the surface it may appear that such discussions do not directly reflect the historical junction at which contemporary Belfast finds itself as a city still experiencing the multiple complex legacies of violent 'ethnonational' conflict, such as unrelenting residential segregation; cyclical bursts of contentious events associated with ceremonial commemorations; and continuing occasional rioting. Yet, clearly here too, urban public space is necessarily an integral part of conflict-transformation strategies, and a respective public discourse of 'shared space' has emerged over the past decade. However, such a discourse, and attempts to give it a physical shape and a performative boost, have reflected the ambiguous social and political meanings buried therein.

Before we analyse these ambiguities through our own research, we explore academic conceptualizations of the idea of *shared public space*. We then develop a particular concept of 'spatial narratives', extending beyond a traditional view of narratives as discursive forms of representation to include enactments/performances and the materiality of urban space. This is followed by a discussion of what we call an 'ontological spatial narrative' of the 'shared city' emerging in our research of urban regeneration in Belfast. In the concluding section we reflect on the capacity of spatial narratives, as an analytical tool, to assist in studying and understanding urban-conflict transformation, and on the extent to which Belfast is becoming a shared city.

From public to shared space: Beyond brief urban encounters

Debates around community cohesion and integration have often been based on a renewed engagement with the 'contact hypothesis' developed by the social psychologist Allport (1954). The hypothesis, broadly speaking, postulates that contact improves intergroup interaction through anxiety reduction. The broad spectrum of work that has been loosely based on it, as McKeown et al. (2012: 83) point out, has 'often assumed that the best way to improve intergroup relations in conflicted societies is to bring groups together in a shared environment'.

However, Valentine's 'reflections on the geographies of encounter' (2008: 325) question a romanticized view of urban encounters as necessarily translating into 'respect for difference' or representing 'meaningful contact' that is capable of changing values beyond the single individual. Instead, she suggests, studies of social interactions in public spaces (e.g. Amin 2002; Holland et al. 2007) show little actual mixing between different groups and their tendency to self-segregate, while streets, as spaces of transit, produce few exchanges and little connection between strangers. One of the reasons for the inability of contact research to explain or predict the failure of encounters in public space in producing more positive attitudes and relationships at group level, Wessel (2009: 12) observes, is that it often 'fails to specify macro-level conditions' and is often bound up in a minimalist conception of space. It differentiates between neighbourhoods, workplaces, schools and so on, but it does not project space as a meaningful object to which people orient their actions. Nor does it consider reserved and dissociated behaviour in urban space (Wessel 2009: 6). This point is important since, as Valentine (2008: 333) additionally argues, mundane encounters 'never take place in a space free from history, material conditions, and power'. Instead, wider societal/group narratives serve as a background against which to interpret encounters or change attitudes, and act as tapestries of feelings (often of injustice, real or imagined) that could serve as an 'emotional bridge' between people's attitudes to individuals and those to wider social groups. Equally importantly, Valentine states (2008), the social psychology literature that draws on the contact hypothesis, even when critical of the latter's limitations, often fails to take into account the multiple intersecting identities (beyond ethnicity) with which we may approach such encounters. For instance, based on observing intergroup behaviour in integrated secondary schools, further education college classes and cross-community groups in Northern Ireland, McKeown et al. (2012: 83) argue that, even in what is deemed 'shared space', Protestant and Catholic young people remain highly segregated in homogenous groups at the individual level. Therefore, they state, 'shared space is not really *shared* in the true sense'. While this work is a welcome critique of the capacity of low-level encounters to produce engagements that can transcend ethnonational conflict, it also demonstrates one of the biggest difficulties with the use of a discourse of 'shared space' in both academic research and public policy: What indeed is the 'true sense' of shared? Whose point of view do we take when we use this notion? And how is it linked to the social meanings and identifications attributed to different spaces?

The above questions are, one may argue, addressed, even if often by implication only, by those researchers concerned with the relationship between 'space, the city and social theory' (Tonkiss 2005). In this type of work, public spaces are understood as shared in the sense that they can serve as common resources and have common effects, but also, intrinsically, as arenas of conflict because the content of citizenship, of collective identities and of spatial rights is worked out through collective struggles over access to, and the meaning and uses of, urban public spaces.

To understand the idea of public space as shared requires an appreciation that different urban spaces have different, though interrelated, functions and meanings, which may bear differently on how sharing is understood and enacted. Tonkiss (2005), for instance, develops an ideal typification of public space, distinguishing between 'the square', 'the street' and 'the café'. 'The square' represents collective belonging and serves as a stage for organized political and social action that enacts civic struggles over meanings, symbolism and collective identities. It is the literal, physical space for the performance of democracy (Parkinson 2012). In this sense, Bryan (2009) argues that by following changes in the identity practices associated with public events in Belfast city centre, we can glean what 'shared space' may be about and how its understanding and practice differ, depending on temporal context. He differentiates between two definitions of the civic through the use of public space: the general use by anybody, including social movements which may oppose or undermine the state and public authorities; and instances of use of space specifically enabled or encouraged by the state/public authorities. By examining how events, such as the St. Patrick's Day Parade, have over time shifted positions from the former to the latter (and are now officially legitimated by Belfast City Council (BCC)), Bryan (2009) suggests that we can understand the nature of the changing relationship between political identities in the city, and see evidence that new types of sharing are taking place. Relatedly, drawing on Lefebvre's ideas about urban, participatory democracy, Nagle (2009: 344) examines 'alternative and even progressive uses for public space' in Belfast, by trade unions, LGBT social movements and carnival practitioners. He suggests that such groups 'can engender new and imaginative uses for segregated and neutral space', leading to a reimagination of the ethnonational city 'as a site of renewed centrality, a place of encounter, [and] an assemblage of difference' (2009: 344).

While the above accounts tend to concentrate on central, representative public spaces, Tonkiss' (2005) categorization of public space includes, just as importantly, 'the street' and 'the café' as the sites of mundane social interactions, social exchange and sociality. As such, she contends, these spaces are basic to everyday modes of being together in public. The street in particular is the space of everyday informal encounters with others, allowing for the 'trivial' sharing of social space, mobility, fluidity and mix of uses which sustain social diversity. It is therefore 'the basic unit of public life in the city'; 'the best and most obvious example of a *shared public space* in which individuals are brought to interact, however minimally, with others' (2005: 68, our emphasis).

The above is not to romanticize or idealize the street as an environment free of power relations or conflicts. Instead, Tonkiss suggests, it is a site of the micropolitics of urban life where individuals exercise their spatial rights and negotiate spatial claims with others. Drawing on the work of Anderson (1990, 1999), she shows that different codes of behaviour, linked to neighbourhood relations, race, age and gender, act as informal means of policing public space, thus showing the true, lived limitations of our spatial rights and how they apply differently to different people. In Tonkiss' account, the real life of public spaces is constituted through various forms of control and exclusion perpetuated both by formal institutions and by the informal fabric of urban everyday life. Sharing, in this sense, is not about lack of conflict since in every city there are contests over the functions, uses and ownership of public space, tied to access, meanings and representations. Shared space here appears as a fluid concept, the meaning and practice of which depend on both broader and specific temporal and spatial contexts, and on the social identities through which it is perceived and practised, because 'Questions relating to how the city gets carved up, and the functions for which its spaces are used, are always tied to questions of who exactly gets to use it' (Tonkiss 2005: 63).

Sharing the city – beyond individual spaces

Linked to a general concern with animating mundane urban spaces, such as streets and neighbourhoods, planners, architects and urban designers have also produced a body of work relevant to our discussion of shared space. Sandercock (2003) advocates designing urban spaces that promote encounters and increase contact. Brand (2009a, 2009b) notes the ways in which different building conventions affect the social use of urban space and can encourage or discourage encounters. Hickey (2014) stresses the link between human perception and the physical dimensions of the built environment with respect to efforts to turn Belfast city centre into 'shared' civic space. She notes the importance of building design and scale, the quality of street layout, the legibility of space and high levels of human activity for inducing feelings of safety, ease, pride and civic ownership.

While such research is notable for emphasizing the mutual constitution between the built environment, social practices and psychological perceptions, Sterrett et al. (2012) argue that approaches to dealing with individual buildings or urban spaces tend to miss the underlying dynamics of urban development and suggest instead that it is necessary to concentrate on questions of urban structure, understood as the pattern of arrangement of the different elements comprising the built environment, and their interrelationship. Similarly, Parkinson (2012: 74) draws attention to the importance of urban connective structures, shaping the broader distribution of networks of opportunities in cities, beyond what individual spaces offer as stages of democratic performance. Urban structure, he suggests, can have exclusive or inclusive effects, and has the capacity to emplace difference and hierarchy. Parkinson gives examples with the Old South African Group Areas Act, which institutionalizes patterns of exclusion through the 'strict spatial separation of suburbs and townships', disconnecting these from urban centres 'not by fences and police but by infrastructure', such as motorways, railways, large drainage systems, power lines and golf courses. Thus planning regulations create barriers to access and encounters, making it less likely for people to feel that they are members of a common public, which is necessary for the functioning of democratic politics.

In this sense, Sheller (2008: 32) stresses that public space only becomes public when people access it, though they can only do so temporarily because it must also always be accessible to others. Applying this point to Belfast, Sterrett et al. (2011) argue that strategic urban planning there has not been sufficiently concerned with how the broader structural development of the city (and the kind of connections and access to individual public spaces that this allows) shapes the very possibility of inclusive civic space. Thus the role of connective structures, beyond central 'representative' space, has remained peripheral to how access to civic space, or the effects of public space itself on the interactions and relationships between different groups, is understood.

The discussion so far suggests that a complex bundle of questions are contained in trying to understand what shared space is. Space, above all, needs to be understood as a relational and meaningful production – a context and a site of power relations of which interactions in public spaces are a product, and in which they play a role. Such interactions are also routinely approached through multiple and intersecting social identities that go beyond generic labels of ethnicity or political affiliations, and express more banal, yet ubiquitous, social differences around age, gender and even sexualities. A relational understanding of urban space also suggests the necessity to be concerned with the effect of wider urban structural development (above and beyond individual spaces and buildings) since the overall plan and structure of a city have the capacity to create social inclusions and exclusions, shaping and representing the conception and practice of a common public (Parkinson 2012). Collectively, the above 'epistemological narratives' suggest that shared space needs to be understood through a combined attention to the meanings (discourses), practices and built materialities of the city.

The city as an assemblage of spatial narratives

Narratives are a ubiquitous feature of social life, most commonly understood as a form of representation. However, Somers and Gibson (1994: 38) argue for an extended understanding of narratives as constitutive of social life which is itself fundamentally 'storied'. 'People construct identities... by locating themselves, or being located within a repertoire of emplotted stories'; they make sense of what happens to them by attempting to 'assemble' and 'integrate these happenings within... narratives'. Similarly, Kleres insists that 'The very nature of human... experience can be conceptualised as essentially narrative in nature (rather than *mediated by* narratives)' (2010: 188, our emphasis). In this view, narratives have ontological functions and dimensions, and they are not simply representational devices, as traditionally understood. In this context, we suggest, one level of understanding contemporary cities comprises multiple ontological narratives – that is, those that urban residents, city officials, policy-makers, community activists or business people, among others, use in order to make sense of their professional and personal experiences, and act in accordance to in their daily lives. These competing and interweaving ontological narratives are often public and institutional, and they reflect attempts by organized groups to impose order and coherence on a fluid and incoherent urban reality. They are, in fact, 'spatial stories' – while told through verbal and written discourse, they are necessarily constructed within, and with reference to, specific places, performed through associated spatial practices, and given form by the material and visual city. Narratives, in other words, comprise discursive, performative and spatial (built environment) elements, manifested in a 'gathering' of ideas and practices. In this sense they create and stabilize places and the social identities associated with them, assigning 'in/appropriate' (Creswell 1996) practices and forms of behaviour for particular places, and giving them a discursive, visual and material outlook. In this manner they give shape and content to sociospatial change, and engage in a narrative field of a struggle over the meaning, outlook and practices of conflict and its transformation.

They tell us what these are about; how they need to be addressed; by whom; and how we should feel about them. They open, suggest or even impose particular directions for urban change at the same time as they resist, preclude or close off others.

Our delineation of a spatial narrative of Belfast as a 'shared city' derives from our work for an Economic and Social Research Council (ESRC) research project entitled Conflict in Cities and the Contested State.[1] It draws largely on one aspect of the overall research, dealing with the interaction between intercommunal divisions and urban regeneration since the 1998 Peace Agreement. In our analysis we draw on interviews and discussions about the future of Belfast (conducted between 2008 and 2013 with planners, urban and regional civil servants, community activists and academics), case studies of major regeneration schemes, and key documents associated with the regeneration of the city.

Belfast: The 'shared city'?

The emergence of an explicit discourse of public space as a vehicle for conflict transformation in Northern Ireland is not surprising, given the fundamentally spatial nature of past and present-day divisions here. Currently, 37 per cent of electoral wards in Northern Ireland are classed as 'single identity' where this is defined as 80 per cent or more of the same religion (Nolan 2013). Significantly, this figure indicates a decline from that of 56 per cent returned during the previous general census (2001), giving Shuttleworth and Lloyd (2013) the grounds to suggest that for the first time since 1971, residential segregation in Northern Ireland has fallen. The troubling aspect of this overall picture, however, is that segregation along ethnonational communal lines remains the prevailing norm among residents of public housing estates, 91 per cent of whom are said to live in 'single-identity' areas (NIHE 2011). Thus territorial divisions in the region are firmly embedded in class relationships.

Such continuing manifestations of division notwithstanding, since the Peace Agreement of 1998 there has been a visible transition in the urban fabric of Belfast, from a city physically scarred by decades of violent conflict to a consumerist city characterized by renovated retail, office and recreational spaces in the centre and along the waterfront. Less clear, however, is whether the new spaces created through such urban renovation and regeneration also facilitate a transformation of the antagonistic ethnonational relationships so long inscribed in the material and social environment of the city. Questions about the social processes behind desegregation and the extent of cross-community interaction in such new urban spaces remain underexplored.

The narrative of Belfast as a 'shared city' emerging from our research is built around, but extends beyond, the notion of 'shared space' which first appeared in the government's good relations strategy, *A Shared Future* (OFMDFM 2005), published in 2005. This document was the result of a commitment in the *Programme for Government, 2001–15* (OFMDFM nd) to put together a strategy for the promotion of community relations. It defined 'shared space' in a somewhat vague and circular fashion as necessary for 'developing and protecting town and city centres as safe and welcoming places', 'creating safe and shared space for meeting, sharing, playing, working and living'; and 'freeing the public realm from threat, aggression and intimidation while allowing for legitimate expression of cultural celebration' (OFMDFM 2005). This definition has been widely criticized for its vagueness and lack of appreciation of the underlying factors shaping the nature and extent of 'sharing' between communities in conflict.

When an OFMDFM consultation document for a strategy entitled *Programme for Cohesion Sharing and Integration*, intended to replace *A Shared Future*, appeared in 2009, it was in the midst of an open conflict between the political parties in the local parliament that had failed to agree even on a common draft of the document. Analyses of this second draft strategy conclude, among other things, that the important goals of 'encouraging shared neighbourhoods, tackling the multiple social issues effecting and entrenching community separation, exclusion and hate' (Todd and Ruane 2009: 4) were not being advanced, and neither was a discussion about how to develop the safety and neutrality of public spaces, suggesting thus a decreasing inclination to even question what might make space 'shared'.

An actual cohesion sharing and integration strategy was never agreed by Parliament and, eventually, in 2013, a new document entitled *Together Building a United Community* (OFMDFM 2013) was released. This strategy now contained an aspiration for removing 'peacewalls' in Northern Ireland by the year 2023, and for the creation of more shared neighbourhoods. Significantly, however, it failed to reach agreement on a number of other questions, including displays of flags, communal marches and the work of the Parades Commission – all fundamentally linked to both territorial claiming of space through organized manifestations of exclusivist identities, and to mundane questions of access and use of different spaces in the city. Furthermore, in the absence of more specific plans to action or negotiate the removal of interface barriers, and the lack of obvious integration of the *Together Building a United Community* strategy into existing public policies, questions have been asked regarding the extent to which the creation of 'shared space' is considered a central policy issue in Northern Ireland (Byrne and Gormley-Heenan 2014).

Different spaces, identities and ideas of sharing

The characteristic vagueness of the above-described policy discourses in defining and strategizing for 'shared space' reflects the ambiguities of broader discourses on conflict and its transformation in Northern Ireland (Komarova 2011). It is also intrinsically linked to the ways in which cultural and political identities are narrated in our data. For instance, exclusive territorialist understandings of 'community' often translate into definitions of 'shared space' that accept the right to express political identities unquestionably, while precluding any scrutiny of the content of such identities, or of the manner of their expression. Here is how a politician, and an ex-government minister, from the DUP talks about how the notion of 'shared space' applies to a road which forms part of the traditional 12th Orange Order Parade in Belfast. This route is contested by nationalists as it passes through what is largely the Catholic-populated Ardoyne area of north Belfast:

> a shared and better future... needs to be applied to open space. And I think that's the most difficult area for Sinn Féin [main nationalist party] because no longer would you be able to say that the Springfield Road is a Nationalist road. If it's a main road – it's a shared road and it should be accessible for people to use. And come a particular day in June [sic], if members of the Orange Order want to use that road they have every right to use it.
>
> (Interview, 20 February 2009)

According to this research respondent, the existing unionist cultural tradition of parading is inalienably related to the nature of unionism, remaining in principle not open for discussion, change or alternative interpretations. Other quotes demonstrate that understandings of 'shared space' are intrinsically related to notions of place – both to the meaningful, experiential and emotional connections tying people to places, and to the histories and accepted uses of place. In the interview excerpt below, a member of a residents' association, representing the same area of Ardoyne referred to above, speaks about his role in an organized protest against the 12th Orange Order Parade passing through the area in 2011:

> it is literally 10 feet away from where my front door is and nobody's come and said to me,... 'do you mind if we walked past here, and we have a union jack and have a paramilitary flag here on the side of my arm, and I am up to my knees in fennian [Catholic] blood, do you mind that?' Nobody has come to my door and said that. So [the protest] is my way of saying, 'I don't particularly like you doing that. I prefer if you spoke to me and had a conversation with me, and we had an agreement as to how you do your culture, or what I think would be acceptable outside my front door.'
>
> (Interview, 3 June 2011)

Clearly in the above the close association between place and cultural identity is seen as a basis for demanding spatial rights, and for contesting those of others. Certainly, in the case of Belfast, discourses of place play a central role and act as a powerful tool, not only in the social construction and contestation of spatial divisions and territoriality but also as manifested in the mundane uses of public space in the city. Indeed, beyond Belfast, discourses of place are often used to the effect of affirming or contesting spatial ideals, and may have the effect of stratifying cities into pools of unequal spatial entitlements (Di Masso 2012).

At the same time, when speaking about the city centre, urban residents participating in our research often expressed views of it as a shared space – 'not...demarcated as nationalist or republican or loyalist and unionist' (roundtable discussion, 26 September 2008) and as 'somewhere where ...well, it doesn't matter who you are when you are using it.... That no one will begin to say who or what you are; it's just like a normal society' (Interview, 11 June 2008). Yet, other research conducted for the Conflict in Cities project questions the type of sharing that such a neutrality achieves. Discussing how mothers of young children from segregated inner-city neighbourhoods use central commercial areas in the city, Smyth and McKnight (2010: 25), for instance, point out that:

> other deeply entrenched social divisions, not least those of class, ethnicity and gender itself, are actively in play in the city centre, shaping the orientations of shoppers both towards each other and their environment...Thus, while the city centre may be a non-sectarian arena which is quite distinct from the surrounding neighbourhoods, its

commercial character ensures that other social divisions continue to shape the everyday orientations of its shoppers...

Similarly, Leonard and McKnight discuss how young people from segregated neighbourhoods in Belfast access and use centrals spaces and organized events in the city, suggesting that for the most part they do so 'in pre-existing, single identity peer groups and view these [events and spaces] as either inclusive or exclusive calling into question the extent to which Belfast's city centre can be viewed as shared space' (2013: 1).

Furthermore, other complexities remain in how people from inner-city neighbourhoods perceive the city centre as a place in the sense of trying to develop some meaningful identification with it. The quote below shines the spotlight on the fuzzy edges of the city centre that act as connection and entry points, and thus their perception as accessible and neutral is equally important:

> to me – in and around the Belfast City Hall at the minute is a shared space. Castle Court is a shared space. Victoria Square is a shared space … But if you say to me in general – do I feel secure in Belfast city centre – No! Because I would feel in and around Castle Street, in and around the Markets – it's not a shared space for me. Going around central train station is not a shared space … It's one community.
>
> (Interview, 21 June 2008)

On a different but related point, urban professionals have argued and demonstrated that it is these types of space at the edges of the city centre which have been adversely affected by fragmenting road infrastructure, blighting of space and residential decline. Such factors have reinforced a high degree of pedestrian disconnection/isolation, cutting off surrounding inner-city neighbourhoods from the centre – physically and psychologically – while ensuring good commuter car access from the more affluent south, east and suburbs of the city (Hackett et al. 2011).

Economically driven regeneration

As alluded to in the previous discussion, 'the shared city' narrative is significantly, if not always openly, influenced by a relationship between peacebuilding and economic development where the imperative of desegregating the city is linked to its economic success and competitiveness. As a civil servant from BCC put it,

[Conflict transformation] is critical to the kind of economic success of the city. It's not about some moral nice thing to do. While I privately might think this is a moral thing to do, it is also about the economic success of the city and the fact that we can attract talent, we can attract investment, we could keep the really good people here.

(Interview, 16 December 2009)

The link between economic development and peace is also contained in the *Programme for Government 2011–15*, which features the task of 'growing a sustainable economy' (OFMDFM nd: 3) as the first of five priorities for the work of government, above and beyond that of developing and 'building a strong and shared community' (OFMDFM nd: 3). Arguably, economic development as the driver of peace has successfully manifested as a 'shared' discourse, expressing a degree of ostensible political consensus. The practical enactment of this discourse has been represented by the completion of a variety of economically driven regeneration projects in the city, the largest and most well known of which is the Titanic Quarter (TQ).

When the 'iceberg-shaped' building of the Titanic Signature Project was erected in the former Harland and Wolff shipyard area of Belfast (the birthplace of the iconic ship *Titanic*), it was not simply the words of its chief executive that the future of Belfast is now about 'innovation-led research and development' (Interview for *Belfast Telegraph*, 1 November 2010) which propounded the argument that economic development drives conflict transformation. Neither was this idea simply contained in the countless press releases, interviews and speeches by representatives of the public-private partnership that presided over turning this part of the city into 'Europe's largest and most exciting waterfront development' (Titanic Quarter Ltd. 2012). The ideological construction that 'peace' is about normalization through neoliberalization of the Northern Ireland economy (with Belfast as its core) has not been a 'simple' matter of discourse but is also driven and manifested in the multitude of activity spaces offered by the TQ itself, including:

a mile of water frontage and a range of investment opportunities . . . : over 7,500 apartments, 900,000 sq. m. of business, education, office and research and development floor space, . . . hotels, restaurants, cafes, bars and other leisure uses.

(Titanic Quarter Ltd. 2012)

Notably, social and affordable housing is not part of this variety of everyday life spaces, and the associated activities and uses, while the connection of TQ to the rest of the city to this day remains favourable to private car use. Thus the sociospatial practice of regeneration, and the visual and material change that comes with it, are just as clear in insinuating the appropriate social activities and practices that can take place in TQ as are any number of eloquent words. What such a narrative spells out is neither simply that the economy is important, nor that Belfast has moved on from the times of violent conflict by relying on economic development, but that there exist appropriate spaces ('iconic' sites) and social classes that visually represent, and are able and willing to partake in, both these transformations of the city. The regeneration of the area, and its turning into a place of consumption, represents and stabilizes spatially the middle-class social identities of its mundane users.

Still, Nolan's statement (2013: 129) that regeneration is hugely important in converting territory that was once 'orange' or 'green' into 'neutral space' rings a certain truth for places, like the TQ, which have been developed in parts of the city that do not encompass existing residential neighbourhoods. By contrast it is exactly the task of physically developing shared urban space in traditional 'orange' and 'green' residential territories that has proved to be the most challenging. Plans to develop shared spaces in territorially contested/segregated parts of Belfast have failed with some regularity, since once tied down to such places in the city, real differences in the understanding of sharing have crystallized.

Examples are easily found, and perhaps none has been more notorious in recent years than the 'failure to launch' the Crumlin Road Gaol and Girdwood Barracks Regeneration Scheme (O'Dowd and Komarova 2011). The planned regeneration is located in north Belfast, among a mosaic of Loyalist and Republican communities, fractured by walls and peacelines. North Belfast was one of the key cockpits of conflict throughout the Troubles, and in the years since the Good Friday Agreement it has seen the most frequent and serious incidents of sectarian unrest and violence, including a number of riots over the passing of the 12th Orange Order Parade through the Nationalist Ardoyne area. Even more poignantly, north Belfast is notorious for deprivation in employment, housing, health and education. Some parts of it are prosperous but the five electoral wards immediately surrounding the regeneration site in question all rank in the top 5 per cent most deprived in Northern Ireland (NISRA 2010). North Belfast remains firmly part of Troubles or 'Interface' Belfast. Territoriality here is particularly heightened in the context of housing, where overcrowded Catholic/Nationalist areas are juxtaposed with unoccupied and rundown spaces in some Protestant/Unionist areas.

The draft masterplan for the Crumlin Road Gaol and Girdwood Barracks Regeneration Scheme was based on an aspiration for 'a transformational *Shared Future* scheme' (e.g. developing shared space) (DSD 2007). It proposed a number of developments comprising a £320 million investment in the 27-acre site over a period of 15 years – an investment unheard of in that part of the city. It had the potential to positively transform the lives of the communities living around it, yet it was most adamantly challenged over its proposals to develop 'shared' residential space. Bluntly put, Protestant communities and politicians vetoed such a development because the overwhelming housing need among Catholics in the area would have meant that any social housing built on the site was to be allocated to people from a Catholic community background. The fears expressed have been of the development site turning into 'Catholic territory', which was seen to be in conflict with the aspirations of the scheme to develop 'shared space'. In May 2012, after long years of political deadlock, a breakthrough was announced regarding the regeneration scheme (BBC 2012). The new plan eschewed the development of shared housing – the most contentious aspect of the scheme – in favour of separate residential developments in different parts of the site, adjacent to the respective Unionist and Nationalist 'communal territories'. Building work on the site has since proceeded accordingly.

Not only does this case study illustrate the lack of a comprehensive approach to place-making on the part of local government administration (Department for Social Development (DSD), to which responsibilities for regeneration are allocated) but it suggests that the administration of the regeneration scheme itself may have been an unwitting accomplice in a *de facto* exercise in territorialism. For instance, from the outset the department delimited the physical boundaries of the site, for which the masterplan was to be developed, failing to define it in relation to other parts of the city and, specifically, to a number of long-existing, but not publicly consulted on, regeneration schemes for adjacent spaces and neighbourhoods. The decision created fixed bureaucratic and spatial boundaries within which the competition for the discursive claiming of territory on the part of communities developed. This was set against the background of a complex and fragmented governance structure in Belfast resulting, as Bradley and Murtagh (2007) have argued, in a lack of clear policy integration, unsatisfactory engagement with segregation and territoriality, and ineffective policy delivery. Thus instead of leading on the process of developing 'shared spaces', the bureaucratic management of regeneration projects has acquiesced with the mutual vetoes of communities over territory in the city.

Furthermore, discursive territorialist competition regarding the regeneration site was consolidated and drawn upon by local political representatives. While politicians may have a vested interest in the development of Belfast as a 'new capitalist city', they are also guided by the imperative not to erode the territorial nature of their own electoral base (the regeneration site being situated in a part of Belfast which is a real patchwork of territorially divided communities). Thus, in the case of this regeneration scheme, local political representatives firmly stood on party-ideological grounds, effectively undermining the original official 'government speak' about a necessity to develop 'shared space' through shared housing on the site. Sinn Féin representatives rejected the validity of the concept of 'shared space', seeing it as sacrificing equality and social justice to intercommunal harmony. DUP representatives claimed to fully support the principles behind 'shared space' but their arguments were constructed in a way that rhetorically undermined the importance of the idea of sharing with respect to residential space.

The fraught development of that regeneration scheme, and the political debates surrounding it, demonstrate succinctly that the unprecedented level of political agreement over the importance and direction of economic development, manifested by an idea of Belfast as 'a new capitalist city' and espoused as a tool for achieving a 'shared city', falls apart once crystallized around those places where ethnonational divisions (on which political affiliation itself is based) take primacy in discourse and action.

Conclusions

We began by discussing the limitations of the 'encounters paradigm', which draws attention to an indispensable aspect of public spaces – their human dimension as spaces for meeting. However, this approach was critiqued for failing to go beyond a minimalist conception of space. Critics suggest instead the need for a broader understanding of space – as a meaningful relational production and, fundamentally, as the stage for the enactment, and the outcome itself, of power relations. Thus 'shared space' needs to be understood as encompassing both organized struggles and mundane competition over the meaning and uses of public space, including questioning the content of collective identities and the manner of their expression, rather than simply the voicing of demands between groups. Collective identities and places are not static phenomena, despite some political and communal discourses to the contrary, yet capturing the more dynamic aspect of the relationship between identity and place has been a challenge for research on 'shared space' in conflict societies. Here the development of 'shared spaces' needs to reflect an understanding of the embeddedness of different types of space into wider urban structures, practices and histories. The idea of sharing and its actual realization would thus apply differently at different times and in different spaces in the city so that their planning and development need to be approached in an integrated manner.

Neither, in the context of studying 'shared space' in divided cities, has Valentine's (2008) focus on the intersectionality of social identities often been adopted by researchers. As she argues, it is necessary to pay more attention to the specific identifications (not just of ethnicity) through which encounters with others are approached, 'and how these encounters are systematically embedded within intersecting grids of [spatialized] power relations' (2008: 332). Smyth and McKnight (2010) are among the few who point out that deeply entrenched social divisions, along the lines of class, ethnicity and gender, are actively at play in Belfast city centre, shaping orientations to everyday encounters and to place while themselves perpetuating 'banal' forms of social division.

In approaching the research on how Belfast may be developing as a 'shared city', we highlighted the usefulness of ontological narratives both as analytical tools and as mediators of urban spatial change. As approaches to understanding the urban, narratives provide ways to consider how different modes of power are exercised through the spatiality of urban life in a city which continues to experience ethnonational divisions. Here we agree with Dovey's (2010: 13) argument that understanding place purely as, and through, discourse (i.e. a matter of written or spoken narration) is itself insufficient because it reduces place to text, 'bypass[ing] the question of ontology and strip[ping] the sense of place of some of its most fertile complications'. We proposed, therefore, a notion of spatial narratives, understood as the assemblage of physical space (the built environment), discourses and practices, in order to better demonstrate that 'Place is at once experienced, structured and discursively constructed' (Dovey 2010: 13), and expose the multiple and complex relationships between power and place.

We use the notion of spatial narratives in a double sense. On the one hand we envisage an analytical tool helping us to address the case study of Belfast's post-conflict transitions and, more broadly, the role of cities in ethnonational conflicts and their transformation. On the other hand, we highlight the ontological significance of narratives. Narratives, we contend, constitute place through the grounding of social identity and have the effects of fixing (albeit temporarily) meanings to places, in different ways, at different scales and with different cumulative effects of power. While our narratives can be seen as representations, when articulated they themselves begin to constitute places, practices and meanings. Moreover, once in existence and 'circulation', narratives become tools for action – the articulation of wider discourses of power that aim to promote or resist sociospatial change.

Finally, we find that a narrative of Belfast as a 'shared city' is discursively appealing because it promises to advance reconciliation and conflict transformation. However, in its discursive prevalence it appears as something of an empty signifier – imbued with various and, at times, contradictory meanings. Its performative dimension is relatively weak and more restricted to central urban spaces. It consists of making communal events more inclusive and removing signs of communal division from central urban space. Yet in many ways this narrative remains aloof to the highly spatialized experiences of division and social deprivation so characteristic of Belfast, and it fails to turn 'new consumerist' spaces into places that are welcoming, meaningful and relevant to marginalized groups in the city.

Note

1. ESRC large grant no. RES-060-25-0015, http://www.conflictincities.org.

References

Allport GW (1954) *The Nature of Prejudice*. Reading, MA: Addison-Wesley.
Amin A (2002) Ethnicity and the Multicultural City: Living with Diversity. *Environment and Planning A* 34(6): 595–980.
Anderson E (1990) *Streetwise: Race, Class and Change in an Urban Community*. Chicago: University of Chicago Press.
Anderson E (1999) *Code of the Street: Decency, Violence and the Moral Life of the Inner City*. New York: W.W. Norton.
Bradley C and Murtagh B (2007) *Good Practice in Local Area Planning in the Context of Promoting Good Relations*. Belfast: BCC.
Brand R (2009a) Written and Unwritten Building Conventions in a Contested City: The Case of Belfast. *Urban Studies* 46(12): 2669–2689.
Brand R (2009b) Urban Artifacts and Social Practices in a Contested City. *Journal of Urban Technology* 16(2–3): 35–60.
Bryan D (2009) Negotiating Civic Space in Belfast or The Tricolour: Here Today, Gone Tomorrow. *Conflict in Cities Working Paper 13*. Available at:

http://www.arct.cam.ac.uk/conflictincities/PDFs/WorkingPaper13_7.1.10.pdf (accessed 3 July 2015).

Byrne J and Gormley-Heenan C (2014) Beyond the Walls: Dismantling Belfast's Conflict Architecture. *City: Analysis of Urban Trends, Culture, Theory, Policy, Action* 18(4–5): 447–454.

Creswell T (1996) *In Place/Out of Place. Geography, Ideology and Transgression.* Minneapolis: University of Minnesota Press.

Di Masso A (2012) Grounding Citizenship: Toward a Political Psychology of Public Space. *Political Psychology* 33(1): 123–143.

Dovey K (2010) *Becoming Places. Urbanism/Architecture/Identity/Power.* Abingdon: Routledge.

DSD (2007) Crumlin Road Gaol and Girdwood Barracks Draft Masterplan. Available at: http://www.dsdni.gov.uk/crumlin-masterplan.pdf (accessed 2 May 2012).

Hackett M, Hill D and Sterrett K (2011) Shared Space. Report, Belfast Conflict Resolution Consortium, Belfast. Available at: http://www.forumbelfast.org/cmsfiles/events/all/shared-space-report.pdf (accessed 3 July 2015).

Hickey R (2014) The Psychological Dimensions of Shared Space in Belfast. *City. Analysis of Urban Trends, Culture, Theory, Policy, Action* 18(4–5): 440–446.

Holland C, Clark A, Katz J and Peace S (2007) Social Interactions in Urban Public Places. York: Joseph Rowntree Foundation. Available at: www.jrf.org.uk/publications/social-interactions-urban-public-places (accessed 3 July 2015).

Home Office (2001) *Community Cohesion. A Report of the Independent Review Team.* London: Home Office.

Kleres J (2010) Emotions and Narrative Analysis: A Methodological Approach. *Journal for the Theory of Social Behaviour* 41(2): 182–202.

Komarova M (2011) Imagining a 'Shared Future': Post-Conflict Discourses on Peace-Building. In Hayward K and O'Donnell C (eds) *Political Discourse and Conflict Resolution. Debating Peace in Northern Ireland.* Abingdon: Routledge, pp. 143–159.

Leonard M and McKnight M (2013) Traditions and Transitions: Teenagers' Perceptions of Parading in Belfast. *Children's Geographies* 13(4): 398–412.

Lownsbrough H and Beunderman J (2007) Equally Spaced? Public Space and Interaction between Diverse Communities. DEMOS: A Report for the Commission for Racial Equality.

McKeown S, Cairns E and Stringer M (2012) Is Shared Space Really Shared? *Shared Space* 12: 83–93.

Nagle J (2009) Sites of Social Centrality and Segregation: Lefebvre in Belfast, a 'Divided City'. *Antipode* 41(2): 326–347.

Nagel C and Staeheli L (2008) Integration and the Negotiation of 'Here' and 'There': The Case of British Arab Activists. *Social and Cultural Geography* 9(4): 415–430.

NIHE (Northern Ireland Housing Executive) (2011) Housing Selection Scheme: Preliminary Consultation Paper. Available at: http://www.nihe.gov.uk/housing_selection_scheme_preliminary_consultation_paper.pdf (accessed 3 July 2015).

NISRA (Northern Ireland Statistics and Research Agency) (2010) Northern Ireland Multiple Deprivation Measure. Belfast: NISRA. Available at: www.nisra.gov .uk/deprivation/archive/Updateof2005Measures/NIMDM_2010_Report.pdf (accessed 3 July 2015).

Nolan P (2013) Northern Ireland Peace Monitoring Report No. 2. Belfast: Community Relations Council. Available at: www.community-relations.org.uk/wp -content/uploads/2012/01/ni-peace-monitoring-report-2013-layout-1.pdf (accessed 3 July 2015).

O'Dowd L and Komarova M (2011) Contesting Territorial Fixity? A Case Study of Regeneration in Belfast. *Urban Studies* 48(10): 2013–2028.

OFMDFM (nd) *Programme for Government 2011–15*. Available at: http://www .northernireland.gov.uk/pfg-2011-2015-final-report.pdf (accessed 7 July 2015).

OFMDFM (2005). *A Shared Future. Policy and Strategic Framework for Good Relations in Northern Ireland*. Available at: http://www.asharedfutureni.gov.uk/index/ consultation-paper.htm (accessed 3 July 2015).

OFMDFM (2013) *Together Building a United Community Strategy*. Available at: http://www.ofmdfmni.gov.uk/together-building-a-united-community -strategy.pdf (accessed 3 July 2015).

Parkinson J (2012) *Democracy and Public Space. The Physical Sites of Democratic Performance*. Oxford: Oxford University Press.

Phillips D, Davis C and Ratcliffe P (2007) British Asian Narratives of Urban Space. *Transactions of the Institute of British Geographers* 32(2): 217–234.

Sandercock L (2003) *Cosmopolis II: Mongrel Cities in the 21st Century*. New York: Continuum.

Sheller M (2008) Mobility Freedom and Public Space. In: Bergmann S and Sager T (eds) *The Ethics of Mobilities: Rethinking Place, Exclusion, Freedom and Environment*. Aldershot: Ashgate, pp. 25–38.

Shuttleworth I and Lloyd C (2013) Moving Apart or Moving Together? A Snapshot of Residential Segregation from the 2011 Census. *Shared Space* 16: 57–70.

Smyth L and McKnight M (2010) The Everyday Dynamics of Belfast's 'Neutral' City Centre: Maternal Perspectives. Conflict in Cities Working Paper 15. Belfast: Queen's University.

Somers M and Gibson G (1994) Reclaiming the Epistemological 'Other': Narrative and the Social Constitution of Identity. In: Calhoun C (ed.) *Social Theory and the Politics of Identity*. Cambridge, MA: Blackwell, pp. 37–99.

Sterrett K, Hackett M and Hill D (2011) Agitating for a Design and Regeneration Agenda in a Post-Conflict City: The Case of Belfast. *The Journal of Architecture* 16(1): 99–119.

Sterrett K, Hackett M and Hill D (2012) The Social Consequences of Broken Urban Structures: A Case Study of Belfast. *Journal of Transport Geography* 21: 49–61.

Titanic Quarter Ltd. (2012) *Titanic Quarter Belfast, Northern Ireland*. Available at: http://www.titanicbelfast.com/ (accessed 17 December 2009).

Todd J and Ruane J (2010) *From 'A Shared Future' to 'Cohesion, Sharing and Integration'. An Analysis of Northern Ireland's Policy Framework Documents*. York: Joseph Rowntree Charitable Trust.

Tonkiss F (2005) *Space, the City and Social Theory. Social Relations and Urban Forms.* Cambridge: Polity.

Valentine G (2008) Living with Difference: Reflections on Geographies of Encounter. *Progress in Human Geography* 32(3): 323–337.

Wessel T (2009) Does Diversity in Urban Space Enhance Intergroup Contact and Tolerance? *Geografiska Annaler: Series B* 91(1): 5–17.

14
Geographies of Crime and Justice in Bosnia and Herzegovina

Zala Volčič and Olivera Simić

Introduction

'Space' has become a keyword in a variety of critical approaches to the study of culture, and the 'spatial turn' – the acknowledgment of the constitutive role of space in social relations – has proved to be a productive one in a range of disciplines across the humanities and social sciences (see e.g. Buck-Morss 1983; Innis 1951; Lefebvre 1991; Soja 1989; Thrift 2002). The identity of spaces is very much connected to the histories which are recounted about them, how those histories are narrated and which interpretation of history becomes dominant. Hundreds of diverse locations across today's Bosnia and Herzegovina (BiH) are associated with the armed conflicts of the the 1990s – these are the spaces that witnessed rape, torture and massacres. Some of these were purpose-built, but many everyday ordinary places were also transformed. For example, schools, sports halls and hotels were reworked for 'extraordinary' purposes into spaces of crime.

As such, the Bosnian landscape is 'loaded' with recent tragic events and it remains a tainted ground. Although such spaces are contested, they are common to post-conflict societies such as Germany, Argentina and Rwanda (Burström and Gelderblom 2011; Hite 2015; Schindel 2012). In this chapter we discuss legacies of spaces where torture and rape were committed during the Bosnian War. We pay particular attention to the questions of contested interpretations about two specific spaces of crime: a spa hotel, Vilina Vlas in Višegrad, and the Court of BiH in Sarajevo (the Court). We are interested here in the ways to acknowledge, and remember, spaces of crime. Can these sites of crime be used as models for community-based education and/or tourist destinations? Can, and how could, these sites heal communities ravaged by war? We are concerned with questions of space, studying how particular spaces were negotiated, and how they shape people's experiences, memories, feelings and interpretations. In studying spaces of crime, we see these two sites as particularly powerful and illustrative in showing how the past is present in places in a variety of ways – materially and symbolically.

Our first case study deals with crime, space and dark tourism. In 2014, acclaimed Bosnian film director Jasmila Žbanić released her movie *For Those Who Can Tell No tales*, which is based on a true story of Kym Vercoe, an Australian actress. Vercoe was faced with 1990s wartime atrocities when she visited a Bosnian town, Višegrad, as a tourist in 2008. Upon a recommendation provided by a tourist guidebook, she stayed at Vilina Vlas, not knowing that it used to be a notorious rape camp during the war, a place where rapes of Bosnian Muslim women were committed by the Serb forces. After finding out about the women who were raped on a mass scale there, Vercoe called and confronted the writer of the BiH tourist guidebook, Tim Clancy, to ask him why he had included and recommended Vilina Vlas. As a direct consequence of the conversations Clancy had with Vercoe, he decided to delete the place from the BiH tourist guidebook. We situate our analysis within a scholarship that has analysed how spaces of crime are left out of the guidebooks – sites of recent violent events that remain unacknowledged or silenced (Kaplan 2011: xviii, 253; Winstone 2010).

The second case focuses on the building where the Court resides. According to Amnesty International, there have been a number of allegations that Bosnian Government forces detained Serb women for rape, torture and sexual abuse in various locations in Sarajevo. Some of the locations cited as having been used for such purposes have been a student hostel 'Mladen Stojanović' in Radićeva Street, premises in Danila Ozme Street, and in the Alipashino Polje and Čengić Vila quarters of Sarajevo (Amnesty International 1993: 12). One such space is also the building where the highest legislative body of BiH is situated – the Court the space that during the war was turned into the Military Court of the Bosnian army. This court served, according to former Serb camp inmates associations and victims, as a torture camp for Serb civilians.

This chapter will critically analyse different historical narratives and competing memories embedded in sites of recent torture and killing, and today's justice (the Court) and tourist leisure (a spa hotel, Vilina Vlas). The different engagements with these places and how they should be physically remembered has become part of ongoing contestation around the interpretation of the Bosnian War and the (re)imagining of the post-war society. Such spaces are fleeting – full of contradictions and paradoxes, contingent and under constant formation and (re)interpretation. They have been turned into spaces of conflicting remembering processes (Logan and Reeves 2008).

This chapter has three major sections. In the first theoretical section, we argue that any political and cultural elements of transformation have a strong spatial dimension which one needs to acknowledge. As Massey (1995: 187) argues, 'the past may be present in the unembodied memories of people, and in the conscious and unconscious constructions of the histories of the place'. Space and place represent crucial organizing frames for the more general way in which we understand the world and make our way around it. We also argue for the greater recognition of the significance of the spatial, and we take space seriously at conceptual levels when analysing memories, crime, trauma and remembering. In the second section the two case studies of Viliana Vlas and the Court are analysed. In the third section we offer some concluding remarks.

Theoretical framework: Space, place and memories

Throughout everyday life in the (national) community, we negotiate space, positioning ourselves socially, physically, morally, politically and metaphorically in relation to others. When confronted with the name of a place, we think first in spatial terms – for example, where do we position Sarajevo? We also conceptualize places as always constructed out of articulations of social relations (e.g. historical connections), which are not only internal to that place but link it to other places, elsewhere.

In this section we are particularly interested in the theoretical arguments about spatial layers of identities and memories of spaces of crime. The concepts of space and place were excluded in early definitions of the terms identity, or memory, and place is still being neglected and often treated as a simple landscape or territory description. The study of place, however, is more than mere description: it is 'an organised world of meaning' (Tuan 1977: 179), and people are 'topophilic' creatures: they are intrinsically bound to specific landscapes (Tuan 1977). Place provides a sense of orientation for both individuals and communities. According to Schama (1995), people feel attached to specific landscapes because these landscapes are identified with specific stories or memories.

Here we are, in particular, following critical cultural geographers, and the work of Massey (1994) when she reconceptualizes 'the spatial' and argues that space is created through social relations and political processes, and at the same time the social is spatially constructed. The identity of space is always being produced and reproduced, and an understanding of spatiality, Massey writes, entails the recognition of 'power geometry'. Power geometry not only is concerned with the issues of who moves and who doesn't but also is about power in relation to the flows and the movement. Those who move freely have power. On the one hand there are groups who are in charge of time–space compression and use it to their advantage (global businessmen, tourists), and on the other hand there are people who are disenfranchised by the power geometry and mobility of other groups (refugees).

For Massey, places do not have single identities but multiple ones. Places are not frozen in time; they are processes. In particular, the stories of 'imagining' nations, regions and also concrete buildings, through remembering the past and through the construction of discourses of belonging, represent a place-based process, and different groups have different power and interests in recreating them. It is important to recognize not only that spaces are imagined and remembered but also how and where they are imagined, with what interests and, importantly, what ideology lies behind them (Massey 1994).

Places also stretch through time. Places as depicted on maps are places caught in a moment; they are slices through time: any claim to establish the identity of a particular place depends upon presenting a particular reading of that history (Massey 1995: 182–192). Lefebvre (1991) claims that the space we occupy, work in, remember and live in is essentially produced by social processes and practices. To Lefebvre, space is not merely a geographical or physical location or commodity but also a political instrument. Each society offers up its own particular space. In understanding these processes of belonging and imagining the (spatial) nation, an acknowledgement of processes operating at a range of scales is significant in the articulations (Mitchell 2003).

The relationship between collective memory and space is discussed by Halbwachs (1980). Spaces are socially constructed in that they are physical sites given meaning that is informed and understood by a group and the 'structures' of its society. Yet the relationship between a space and the group that gives it meaning is rather cyclical, where the space influences the group's meanings as it 'becomes enclosed within the framework it has built' (Halbwachs 1980: 130). Halbwachs (1980: 140) thus states that collective memory always exists within a 'spatial framework' because it gives it a medium for preservation.

A rich body of literature addresses how to ethically acknowledge and remember the spaces of crime (Lisle 2004; Schwenkel 2006; Timm Knudsen 2011). On the one hand, scholars ask how to retain their historical and commemorative significance, and how to avoid becoming a site of entertainment that is largely detached from the war (Volcic et al. 2013). The transformation of the spaces of crime into sites for touristic consumption, for example, raises serious questions about their potential trivialization and historical detachment. Dark tourism scholars argue that when painful places get branded, trauma and pain become part of the brand image, as seen in tourism to New York's Ground Zero (Lisle 2004; Sturken 2007), to Auschwitz in Poland (Cole 1999), war tourism in Vietnam (Schwenkel 2006), genocide tourism in Rwanda (Joachim 2012), the numerous museums and memorials across Berlin and, recently, the memorializing of Srebrenica, the site of the act of genocide in BiH (Simić 2008, 2009). Volcic et al. (2013) show how after the violent wars of the 1990s, BiH is heavily promoting its tourism industry with the capital city of Sarajevo as the prime destination. Today Sarajevo faces many dilemmas, including how to remember the 1990s trauma and how to represent the city to citizens and foreign visitors. Volcic et al. (2013) show how Sarajevo (and particular spaces where atrocities happened) is transformed into a tourist attraction where the trauma of the past is being sold, while promising some authentic path towards understanding it.

On the other hand, some tourism scholars point to the educational benefits of conflict-related tourism (Bell 2009; Causevic 2010; Miles 2014), and focus on the importance of mourning about these places that helps to affirm the complexity and richness of witness and testimony as ongoing tasks (Chouliaraki 2012; Cole 1999). In that sense it becomes important to question what has and has not been marked as a site of collective memory of a past atrocity. Why are some sites erased from public consciousness and denied remembrance?

Logan and Reeves (2008) critically evaluate the potentials and pitfalls of different sites of crime. They explore massacre and genocide sites, wartime internment camps, civil and political prisons, and places of 'benevolent internment' (asylums for the insane, outcasts and migrants) as sites that represent a legacy of painful periods in a local or national community history. They analyse how these sites can be, and are being, interpreted through 'planning and management interventions' (2008: 1).

Now and then: Spaces of crime

We want to focus next on the lasting influence of 'spatial frameworks' on collective memory. The two sites of crime that are analysed in this study are the spa hotel Vilina Vlas in Višegrad and the Court in Sarajevo. We want to show different interpretations of the identity of spaces of crime, based and dependent on the different ethnonational position of the involved groups. Moreover, each of these contesting interpretations depends on the mobilization of a particular reading of the buildings' past. And these conflicting interpretations of the past serve to legitimate a particular understanding of the present. At issue are competing histories of the present, wielded as arguments about interpretations of places. The past helps to shape the present, but it is a two-way process, as Massey argues, because presences of the past are multivocal (1995: 187).

Vilina Vlas: Space of leisure (and rape)

Vilina Vlas (Fairy's Hair) was built in 1982 and is situated at the outskirts of Višegrad, a town in eastern BiH, on the banks of the River Drina. Before the war almost 60 per cent of Višegrad's 20,000 residents were Bosniacs. In 2009 only a handful of survivors had returned to what is now a predominantly Serb town (Irwin, 2009). During the war the Bosnian Serb army and its paramilitary forces committed horrendous atrocities. These included burning Bosniak civilians alive and slaughtering hundreds of men, women and children, and throwing them over the Drina Bridge into the river (Irwin 2009).

Before the war broke out in 1992, the hotel was a popular resort for local and foreign tourists alike. During the war the spa was turned into one of 'the most infamous' (Mojzes 2011: 186) and 'the biggest' (Ahmetašević 2011) rape camps in BiH, where hundreds of Bosniak women were raped by Serb forces (Vulliamy 2012: 106). After the war, Vilina Vlas was renovated and it returned to its pre-war function as a mid-grade hotel. It is mostly visited by foreign tourists who want to see the ancient architectural beauties and explore the historical heritage of the region, such as the bridge over Drina made famous by Andrić's novel.[1] Those who visit Višegrad and the bridge will often be invited, either by word of mouth or tourist guides, to visit the once-popular Vilina Vlas. According to one of the BiH tourist guides, 'a tourist postcard of Višegrad is unimaginable without the Višegrad spa [Vilina Vlas], located 5 km north of the town' (BH Tour, 2012).

Following both the recommendations of friends in Belgrade and the advice of a BiH tourist guidebook written by Clancy, Vercoe chose Višegrad and Vilina Vlas as her tourist destination in 2008. Although she stated that she was immersed in learning about the recent dark past of BiH (Robinson 2012), Vercoe had no idea that the hotel had been used as a detention camp for Bosniak women, who were brutally raped and tortured by Serb forces during the war. Most of the women were killed. Their bodies are suspected to be buried below the river bed of the beautiful Drina, a mere stone's throw away from where Vercoe spent the night. After learning the truth about Vilina Vlas following her stay, Vercoe was understandably disgusted and traumatized. Sarajevo director Jasmila Žbanić made a film, *For Those Who Can Tell No Tales*, based on Vercoe's experiences, using the story as a way to break the silence surrounding the rapes, and spaces of crime, both in Višegrad and across BiH.

Since public silence still engulfs Vilina Vlas and Višegrad town, Žbanić opens the story of war rape and particular spaces of crime, while commemorating the female victims. She demands acknowledgement for the women who suffered wartime rape, and she breaks the deafening silence about sexual violence committed *en masse* in the Bosnian War. The film powerfully pays tribute to the city, and to the memories of the Višegrad massacres. Through Vercoe's story, the viewers become haunted by spatial evocations of the atrocities, and by the questions of why the guidebook, or the town itself, made no mention of the recent tragic events (Robinson 2012).

Žbanić uses various conventions traditionally associated with the (historical) documentation and recording of time, such as a flashback, to render an event forcibly effaced into a non-event back into an event. The film also uses narrative conventions of horror films to elicit the viscerality of what Vercoe is feeling – her repeated horror of the realization that she has slept in the 'same bed' and shared space where the rapes occurred, and crossed the bridge of the massacres, suggests more a discomfort with her proximity to the spatial atrocity than the fact of the atrocity itself, as if the spatial proximity meant acknowledging that the violence could have happened to her. The film's attention is focused on the crucial process of coming to terms with atrocious historical events in the absence of official memorialization as a place seeks to move forward, and on the alternative modes of memorialization that can occur.

Profoundly affected by her experience, Vercoe created her own solo theatre show, *Seven Kilometres North East: Performance on Geography, Tourism and Crime*, exploring the uncomfortable links between the spaces of crime, tourism, silence and the (im)possibility of being an innocent tourist in a post-war context. She brought together elements of a travelogue built from verbatim notes of her interviews, guidebooks, war-crime reports, quotes from Ivo Andrić and video images filmed by the artist herself. The role of the theatre piece according to Vercoe was to represent some sort of a living memorial to the victims.[2]

Both in the theatre performance and in the film, Vercoe is depicted as being touched by the fate of the women and the total lack of a public memorial to commemorate them. She returns to Višegrad in December of the same year but cannot find any locals who will admit knowing anything about a rape camp. Everyone, from the suspicious local police chief to the man who chats with her in a bar, denies that any such thing happened at the hotel, or that there was any ethnic cleansing in the first place. Vercoe takes time to wander around the city, this time fully aware of atrocities, and takes issue with the silence and the clean transformation of the hotel space in particular: 'They changed the sheets, washed the blood off the walls, vacuumed the carpets and reopened it as a hotel. In a just world you simply don't do that' (Crittenden 2012).

Both artistic productions – Vercoe's theatre play and Žbanić's film – focus on the importance of space, and ruminate on how the past connects to the present, and how places can sometimes be permanently marked by something horrific that happened there. They both opened a space for revisiting the past and places that remained if not forgotten then engulfed in silence. In the film, often the camera just beautifully pans from a river landscape, onto the bridge, back to remain on Vercoe's face, creating a strong emotional effect. The landscape itself feels so loaded, and the film's aesthetics allows one to start to understand the landscape as witness.[3] In an interview for an Australian newspaper, Vercoe specifically claimed:

> I never had any real urge to go to visit Srebrenica or anything like that – I just wanted to go and see the country and have a great time, which I did. And so I just couldn't comprehend it. I just couldn't believe that somewhere space would be open as a hotel that had been used in that way, and recommended in a guidebook.
>
> (Robinson 2012)

Vercoe contacted the writer of the BiH tourist guidebook, Clancy, to ask him why he included Vilina Vlas, and even recommended it as a nice place to stay. As a direct consequence of their conversations, Clancy decided to delete the spa from the BiH travel guidebook. In his words, 'After talking to Kym, I had two options – either I wrote about Vilina Vlas and explained in depth what happened there – or not, and delete it' (Interview, Clancy, 11 February 2014). For him it was for a strictly personal reason that he erased the hotel from the second edition of his guidebook. His decision was a consequence of his meeting with Vercoe for a long coffee and having a conversation with her about Vilina Vlas:

The reason I took out Vilina Vlas is a direct result of my con-
versation(s) with Kym. She DOES and DID deserve to know what
happened there if I am recommending it. I didn't know but I recommended it. Now knowing the brutal details of Vilina Vlas and how
the community deals (or does not deal) with what happened there
made the choice very easy – I took it out. If I were to leave it in, I
would have to speak of what happened and I personally don't feel
that a travel guidebook is necessarily the place for that – there are
many, many other books and venues that deal with the who, what,
where, why, and how of the Bosnian war.

(Personal communication, Kym Vercoe, 29 January 2014)

The BiH guidebook that ultimately led her to the site of the crimes was
supposed to fit with the 'good time' sensibility – a paradox at best. One
wants to go to BiH to have a good time, but not to confront the horrific
legacy of recent war wounds. Yet is this possible at all? To have a good
time, free from the unsettling reminders of tragedy in a post-conflict
society, is a tricky goal. For such a guidebook to be truly accurate, would
it not have to include all places of crimes and atrocities by all sides in the
wars because 'all Bosnian citizens suffered greatly in the war' (Nettelfield
2010: 188)? Thus if one includes all sites of crimes, a BiH tourist guidebook could easily turn into a war-crime atrocities guidebook since most
of the country was soaked in blood, and many of its hotels, motels
and other public spaces served as killing centres during the war. On the
other hand, these buildings are not harmless – dealing with them in a
disrespectful and ignorant way would trivialize, and affirm the lack of
acknowledgment for, the crimes that happened there.

All of this raises the larger question about the role of the guidebook
in introducing spaces of crime to tourists. As Clancy states, 'It's not my
responsibility to write about ethnic cleansing . . . in a tourist book. I want
BiH and its communities to move forward' (Interview, Clancy, 11 February 2014). He describes Višegrad as a place that is still marked by its
recent history and a 'culture of silence' (Interview, 11 February 2014)
where, as Vercoe puts it, everything is 'hushed down' but 'certainly all
but forgotten' (BiHbloggen 2013). Even the famous cultural heritage –
the bridge over the River Drina – has taken on another meaning. According to Vercoe, the famous cultural heritage, this incredible space, has
become 'a massive monument to genocide' (Silverstein 2013), a situation that left a lasting impression for her: 'as far as I am concerned I will
never cross that bridge again' (Vercoe 2013). It is not possible to relegate
the realm of war atrocity or dark tourism to only selected locations, such

as Srebrenica,[4] and to imagine that the rest of the landscape is somehow innocent and free from the taint of atrocity.

The Court: Space of justice (and crime)

The highest court of justice is a site with rich cultural history that encapsulates multiple memories. It was established in 2000 as a small institution employing only 15 local judges. In the following years it expanded its jurisdiction and employees, which had increased to 245 staff by the end of 2012.[5] The War Crimes Chamber of the Court, which began its work in March 2005, investigates and prosecutes people allegedly involved in serious violations of international law during the 1992–1995 conflict. Over the last few years the Court has invested efforts in its outreach programme to inform the residents of the country but also the international community about its work. It hosts regular visits by local and international students and researchers, and it publishes a brochure, which contain some basic information about the structure and activities of the Court. The brochure is printed in the local language and English on a high-quality paper, enriched with colourful photographs of the building both before and after its reconstruction (Court of BiH 2012).

While there is a section entitled 'The Court through images', followed by the subsection 'The Court building before reconstruction', there is no narration next to the several images of the damaged grey building ricocheted with bullets. After six such photographs that show the long-forgotten and abandoned building, the next photograph is of todays' modern Court building, freshly painted in bright red. The brochure omits to trace the history of the building back to its original construction or to the events that led to its post-war reconstruction. The narrative of the building and its inhabitants is missing and there is no attempt to expose the interrelation between history, law, place and time, which in the case of this building is vital (Court of BiH 2012).

During the war the Court buildings served as a Bosnian military detention centre called 'Viktor Bubanj'. Before the war the buildings were army barracks in which thousands of young men from across the former Yugoslavia were housed while serving in the army, which was compulsory for all men above the age of 18. The duty of military service was regarded as a rite of passage to manhood and most men grew up with the anticipation of serving their time. At the beginning of the war in 1992, the barracks were transformed and reworked into the mock court of the Bosnian army where Serb civilians, primarily from Sarajevo, were detained, tortured and often killed.

The former Serb detainees strongly objected to the choice of the building where the Court is sited because of its history as a detention centre (ICTJ and Ivanišević 2008: 33). For years they have been trying to mark the space by installing a memorial plaque that would acknowledge the site in which the highest court in the country resides in a former torture centre. According to the association of former camp inmates in RS, around 500 Serbs were killed during the war on today's Court premises (Index 2003; SRNA 2013). Although there is no political will to publically acknowledge that the building and the halls have a dark past, in 2011 the first arrests for the crimes committed in the Court's own premises 20 years ago were made. In the first case, of *Muderizović et al.*, a group of Bosnian military detention centre personnel were indicted for war crimes against Serb civilians (Indictment in the case of Besim Muderizović et al. 2011). They were charged, as members of the joint criminal enterprise, with taking part in the establishment and maintenance of the system of abuse of Serb civilians between June and late November 1992. According to the prosecutor, more than 150 Bosnian Serb civilians were illegally detained in the prison in which they were severely beaten and forced to work under conditions in which many of them died (BIRN BiH 2011).

Petar Čajević, one of the witnesses in the case, is a retired pensioner who is still searching for the bones of his son, who was tortured and killed in Sarajevo's 'Viktor Bubanj' detention centre in July 1992 (Dević 2012; SRNA 2013). In the prison, civilians – both women and men – were brought and charged, without witnesses and evidence being presented or arguments being made on either side. They were accused of espionage, the illegal possession of weapons, or membership of the Serb political party. The sentences they received ranged from ten years in prison to the death penalty (Durmanović 2011). The survivors reported severe conditions in which they were held and that detainees died from beatings, hunger, cold, and lung and other infections (Novosti 2006). According to Strahinja Živak, who spent two-and-a-half years in prison, there were 33 women, some pregnant at the time, who were raped by the prison guards (Živak 1999). Milka Lonco, one of the female detainees, reported the horrific conditions and torture that prisoners were subjected to:

> One night was particularly terrible. They killed someone. We heard screaming and whaling through the corridor ... In the morning when we were taken to the toilet someone said, 'someone was killed here last night'.
>
> (Učanbarlić 2013)

The witnesses, many of them former victims of torture themselves, have reported on the retraumatization that testifying triggers while being present in the same space where they were once detained. The necessity of their physical presence before the Court revives the traumatic experience they survived. Radojka Pandurević, a former detainee, recalls her feelings when entering the Court to testify:

> I have recently been invited to testify before the Court of BiH. Once I entered into the building, I felt sick. I felt weak and fear of thinking that here in this building where 'the justice' is served the horrific crimes were committed against Serbs; that in this very building was torture centre for Serbs and that the Court is oblivious to that fact... that I could meet people in its halls who tortured us... that judge in the case I have to testify and stand before is Davorin Jukić [the same judge who presided in her 'kangaroo' trial during the war].
>
> (Vujević, 2010)

Pandurević was confronted with an almost forced encounter with her horror, being in the very same space in which she experienced brutal interrogation and trial without any legal basis. It is obvious that the Court is forever going to be a defining place in her life that will force her to remember past atrocities. As Cook (2006: 296) writes in relation to sites of genocide in Rwanda, many places in which atrocities occurred attain 'special status' in their preservation and memorializing as means by which to 'reveal the truth' about the context, content and extent of lethal violence at the site. These sites can take on 'place specific biographies', evolving as the social, physical and built environment undergoes transformation (Portelli 1984: 287). As Hartley argues (2014: 16), some of these places are memorialized and formally acknowledged in 'official histories', while others are known to people but are not documented officially by government.

Contrary to the BiH Government, which shies away from acknowledging publically and allowing victims to commemorate the site, the South African Government openly and publically commemorates a site that was once a detention centre and today serves as the Constitutional Court. Constitution Hill in South Africa was originally built for military use and later used as a prison through which gates passed various prisoners, including political opponents to the apartheid regime, such as Nelson Mandela and Mahatma Ghandi. South Africa admits its history of injustice during the apartheid struggle, and it acknowledges that in today's South Africa the highest court, the Constitutional

Court in Johannesburg, is situated in the premises of a prison (Ekala Eco Tours nd).

The prison and old fort have partially been transformed into a living museum, and the exhibition, based on extensive research with ex-prisoners, tells a story of those incarcerated there. In Le Roux's work on the architecture of the court, he argues that the court is a rehabilitated prison site that carries the traces/memories of 'trauma and exclusion' of the past (Le Roux 2004: 59). The Constitutional Court's designers purposefully draw visitor attention to these traces of the past in the site itself and to its history that is marked by particular organization of spaces through law (Van Marle et al. 2012: 567). The Court of BiH as a symbol of justice and order is also a space of memory, of how we remember the past and what it means to our present and future. Perhaps because of its identification with the dominant cohort of the population that is now in control of the landscape that belongs to the 'perpetrator' side in the detention centre, there is reluctance to officially acknowledge the difficult historical footprint of the building, and to allow the memorial which will provide a visual that 'condenses a complex narrative into simple symbol' (Buckley-Zisteland Schäfer 2014: 4) of a public physical site.

Conclusion

In this chapter we have attempted to offer some arguments about landscapes of violence and tragedy. Our examples show intense controversies over spaces. Foote's (1997) work is helpful for our conclusions when he highlights four major outcomes for places associated with tragic events. The most common is the process of rectification, which can lead to the designation of a site and eventually pave the way to sanctification. On the other end of his proposed continuum is the process of obliteration – a frequent situation in places that experienced acts of violence and tragedies now forgotten. The memorialization of various layers of history at both sites is necessary since 'the new space is always haunted by the old' (Van Marle et al. 2012: 567) and carries within it the narratives of the past. As we have shown, and other scholars have highlighted, in the eyes of survivors and witnesses the visual-spatial component of memory is significant: 'places are the trigger and, at the same time, the setting of memory' (Cappalletto 2003: 254).

Ochman (2010: 525) writes that working outside this dichotomy would not be to the end of 'moral indifference' but rather would embrace the 'multiplicity' inherent in collective memory. Movement towards this alternative mode of acknowledging spatial memory that embraces complexity does not seem to be accessible in BiH because the sites of trauma are not even acknowledged. In their introduction to *Places of Public Memory*, Dickson et al. (2010: 7) point to the significant relationship between collective memory and space when they state:

> That is, groups tell their pasts to themselves and others as ways of understanding, valorizing, justifying, excusing, or subverting conditions or beliefs of their current moment ... it is to suggest that groups talk about some events of their histories more than others, glamorize some individuals more than others, and present some actions but not others as 'instructive' for the future.

We suggest that just as collective memory is continually negotiated, post-war countries need to address the long-term consequences of creating sites of collective memory in living with the past in the present. By recognizing these spaces, communities can also learn how to accept responsibility for the past and establish the possibility of reconciliation and justice.

Notes

1. The bridge was built in 1571 as an endowment of the Grand Vizier Mehmed-paša Sokolović. Today it is on the UNESCO's World Heritage List.
2. On 22 March 2014 the theatre show was seen by one of us (ZV), who also participated in a question-and-answer session afterwards.
3. For more about the film, and the history of crimes in Višegrad, see Simić and Volcic (2014).
4. Srebrenica is an infamous site of genocide in BiH. In July 1996 approximately 8,000 men and boys were killed in the course of four days.
5. The Court employs both local and international judges and serves as a model of a hybrid war-crimes tribunal.

References

Ahmetašević N (2011) Biggest Bosnia Rape Camp: First Indictment. *Radio Netherlands Worldwide*, 23 November. Available at: www.rnw.nl/international -justice/article/biggest-bosnia-rape-camp-first-indictment (accessed 23 March 2015).

Amnesty International (1993) Bosnia-Herzegovina: Rape and Sexual Abuse by Armed Forces. Available at: www.amnesty.org/en/documents/EUR63/001/ 1993/en/ (accessed 23 March 2015).

Bell C (2009) Sarajevska Zima: A Festival Amid War Debris in Sarajevo, Bosnia-Herzegovina. *Space and Culture* 12(1): 136–142.

BH Tour (2012) Visegrad. Available at: http://bhtour.ba/en/visegrad.html (accessed 23 March 2015).

BiHbloggen (2013) Kym Vercoe: During the Filming in Visegrad We Did Not Dare to Say Jasmila's Name. *BiHbloggen*, 27 September. Available at: https://bosnienbloggen.wordpress.com/2013/09/27/kym-vercoe-during-the-filming-in-visegrad-we-did-not-dare-say-jasmilas-name/ (accessed 23 March 2015).

BIRN (Balkan Investigative Reporting Network) BiH (2011) MederizoviciAvdovic: U istrazijos 25 osumnjicenih [Muderizovic and Avdovic: 25 More Suspects under Investigation]. *Justice Report*, 21 October. Available at: www.justice-report.com/en/articles/muderizovic-and-avdovic-25-more-suspects-under-investigation (accessed 23 March 2015).

Buckley-Zistel S and Schäfer S (2014) Introduction Memorials in Times of Transition. In: Buckley-Zistel S and Schäfer S (eds) *Memorials in Times of Transition*. Cambridge: Intersentia, pp. 1–27.

Buck-Morss S (1983) Benjamin's Passagen-Werk: Redeeming Mass Culture for the Revolution. *New German Critique* 29: 211–240.

Burström M and Gelderblom B (2011) Dealing with Difficult Heritage: The Case of Bückeberg, Site of the Third Reich Harvest Festival. *Journal of Social Archaeology* 11(3): 266–282.

Cappalletto F (2003) Long-Term Memory of Extreme Events: From Autobiography to History. *Journal of the Royal Anthropological Institute* 9(2): 241–260.

Causevic S (2010) Tourism Which Erases Borders: An Introspection into Bosnia and Herzegovina. In: MoufakkirO and Kelly I (eds) *Tourism, Progress and Peace*. Cambridge: CAB International, pp. 48–64.

Chouliaraki L (2012) The Theatricality of Humanitarianism: A Critique of Celebrity Advocacy. *Communication and Cultural Studies* 9(1): 1–21.

Cole T (1999) *Selling the Holocaust: From Auschwitz to Schindler*. New York: Routledge.

Cook SE (2006) The Politics of Preservation in Rwanda. In: Cook SE (ed.) *Genocide in Cambodia and Rwanda: New Perspectives*. New Brunswick: Transaction, pp. 293–311.

Court of BiH (2012) Brochure of the Court of BiH, 4th edn. Available at: http://www.sudbih.gov.ba/files/docs/brosura/brosura_eng.pdf (accessed 23 March 2015).

Crittenden S (2012) Building a Bridge on the Drina. *The Global Mail*, 12 April. Available at: www.theglobalmail.org/feature/building-a-bridge-on-the-drina/181/ (accessed 23 March 2015).

Dickson G, C Blair and BL. Ott (2010) *Places of Public Memory: The Rhetoric of Museums and Memorials*. Tuscaloosa: The University of Alabama Press.

Dević R (2012) Dvije decenije od ocaja i cutanja tuzilaca [Two Decades of Despair and Silent Prosecutors]. *GlasSrpske*, 28 December [online]. Available at: www.glassrpske.com/drustvo/panorama/Dvije-decenije-ocaja-i-cutanja-tuzilaca/lat/104681.html (accessed 23 March 2015).

Durmanović S (2011) Logor 'Viktor Bubanj'- istraga nekažnjenih zločina [The Camp 'Viktor Drum' – Unpunished Crime Investigation].*Novi Reporter*, 22 November. Available at: www.nspm.rs/istina-i-pomirenje-na-ex-yu-prostorima/logor-viktor-bubanj-istraga-nekaznjenih-zlocina.html?alphabet=l (accessed 23 March 2015).

Ekala Eco Tours (nd) Old Fort/Constitution Hill – Constitutional Court. Available at: www.ekalatours.com/constitution-hill.php (accessed 23 March 2015).

Foote KE (1997) *Shadowed Ground: America's Landscapes of Violence and Tragedy.* Austin: University of Texas Press.

Halbwachs M (1980) *The Collective Memory.* Trans. Ditter Jr. FJ and Yazdin V. New York: Harper Colophon.

Hartley R (2014) Signifying the Place of Unforgettable Memory: Atrocity and Trauma in a Post-conflict Landscape. *Anthropology Faculty Publications*, University of Nebraska: 1–42.

Hite K (2015) Empathic Unsettlement and the Outsider within Argentine Spaces of Memory. *Memory Studies* 8(1): 38–48.

ICTJ (International Center for Transitional Justice) and Ivanišević B (2008) The War Crimes Chamber in Bosnia and Herzegovina: From Hybrid to Domestic Court. Available at: www.ictj.org/publication/war-crimes-chamber-bosnia-and -herzegovina-hybrid-domestic-court (accessed 23 March 2015).

Index (2003) Policija u Sarajevu sprečila upad i zgradu suda BiH [Police in Sarajevo Prevent Intrusion into the Court Building]. *Index*, 19 September. Available at: www.index.hr/vijesti/clanak/policija-u-sarajevu-sprijecila-upad-u -zgradu-suda-bih/159903.aspx?mobile=false (accessed 23 March 2015).

Indictment in the case of Besim Muderizović et al. (2011) S1 1 K 008241 11 Kro. Available at: http://www.sudbih.gov.ba/index.php?opcija=predmeti &id=648&jezik=h (accessed 8 July 2015).

Innis HA (1951) *The Bias of Communication.* Toronto: University of Toronto Press.

Irwin R (2009) Visegrad in Denial over Grisly Past.*Institute for War and Peace Reporting*, 24 February. Available at: http://iwpr.net/report-news/visegrad -denial-over-grisly-past (accessed 23 March 2015).

Joachim N (2012) *Investigating the Challenges of Promoting Dark Tourism in Rwanda.* Bachelor's thesis, Rwanda Tourism University College, Kigali, Rwanda.

Kaplan BA (2011) *Landscapes of Holocaust Postmemory.* New York: Routledge.

Lefebvre H (1991) *The Production of Space.*Transl. D Nicholson-Smith. Oxford: Blackwell.

Le Roux W (2004) Bridges, Clearings and Labyrinths: The Architectural Framing of Post-Apartheid Constitutionalism. *SA Public Law* 19(3): 629–664.

Lisle D (2004) Gazing at Ground Zero: Tourism, Voyeurism and Spectacle. *Journal for Cultural Research* 8(1): 3–21.

Logan W and Reeves K (eds) (2008) *Places of Pain and Shame: Dealing with 'Difficult heritage'.* Abingdon: Routledge.

Massey D (1994) *Space, Place, and Gender.* Minneapolis: University of Minnesota Press.

Massey D (1995) Places and Their Pasts. *History Workshop Journal* 39(1): 182–192.

Miles S (2014) Battlefield Sites as Dark Tourism Attraction: An Analysis of Experience. *Journal of Heritage Tourism* 9(2): 134–147.

Mitchell D (2003)*The Right to the City: Social Justice and the Fight for Public Space.* New York: Guilford.

Mojzes P (2011) *Balkan Genocides: Holocaust and Ethnic Cleansing in the Twentieth Century.* Lanham: Rowman & Littlefield Publishers.

Nettelfield LJ (2010) *Courting Democracy in Bosnia and Herzegovina: The Hague Tribunal's Impact in a Postwar State.* Cambridge: Cambridge University Press.

Novosti V (2006) Ubistva zbog pesme [Killings Songs]. *Hoboctn Online*, 8 July. Available at: www.novosti.rs/dodatni_sadrzaj/clanci.119.html:277862-Ubistva -zbog-pesme (accessed 23 March 2015).

Ochman E (2010) Soviet War Memorials and the Re-construction of National and Local Identities in Post-Communist Poland. *Nationalities Papers* 38(4): 509–530.

Portelli A (1984) The Massacre at the Fosse Ardeatine: History, Myth, Ritual and Symbol. In: Hodgkin K and Radstone S (eds) *Contested Pasts: The Politics of Memory*. London: Routledge.

Robinson N (2012) Hotel of Horrors Finds a Voice. *The Australian*, 18 June. Available at: www.theaustralian.com.au/arts/hotel-of-horrors-finds -a-voice/story-e6frg8n6-1226397980664 (accessed 23 March 2015).

Schama S (1995) *Landscape and Memory*. London: Harper Collins.

Schindel E (2012) 'Now the Neighbors Lose Their Fear': Restoring the Social Network around Former Sites of Terror in Argentina. *The International Journal of Transitional Justice* 6(3): 467–485.

Schwenke C (2006) Recombinant History: Transnational Practices of Memory and Knowledge Production in Contemporary Vietnam. *Cultural Anthropology* 21(1): 3–30.

Silverstein M (2013) TIFF Interview: Jasmila Zbanic and Kym Vercoe – Director and Co-Writers of *For Who Can Tell No Tales*. In: *Indiewire*. Available at: http: //blogs.indiewire.com/womenandhollywood/tiff-interview-jasmila-zbanic -and-kym-vercoe-director-and-co-writers-of-for-those-who-can-tell-no-tales (accessed 29 May 2015).

Simić O (2008) A Tour to a Site of Genocide: Mothers, Bones and Borders. *Journal of International Women's Studies* 9(3): 320–330.

Simić O (2009) Remembering, Visiting and Placing the Dead: Law, Authority and Genocide in Srebrenica. *Law Text Culture* 13: 273–310.

Simić O and Volcic Z (2014) In the Land of Wartime Rape: Bosnia, Art and Reparation. *Griffith Journal of Law & Human Dignity* 2(2): 377–401.

Soja EW (1989) *Postmodern Geographies: The Reassertion of Space in Critical Social Theory*. London: Verso.

SRNA (2013) Bezuspješno pokušavao da spase sina iz kasarne Viktor Bubanj [Unsuccessfully Trying to Save His Son from the Barracks Viktor Drum]. *Nezavisne Novine*, 8 March. Available at: www.nezavisne.com/ novosti/bih/Bezuspjesno-pokusavao-da-spase-sina-iz-kasarne-Viktor-Bubanj- 183211.html (accessed 23 March 2015).

Sturken M (2007) *Tourists of History: Memory, Kitsch, and Consumerism from Oklahoma City to Ground Zero*. Durham: Duke University Press.

Thrift N (2002) The Future of Geography. *Geoforum* 33(3): 291–298.

Timm Knudsen B (2011) Thana tourism: Witnessing Difficult Pasts. *Tourist Studies* 11(1): 55–72.

Tuan YF (1977) *Space and Place: The Perspective of Experience*. Minneapolis: University of Minnesota.

Učanbarlić S (2013) Strasnanoc u 'Viktor Bubnju' (Horrific Night in Former Sarajevo Army Barracks). *Justice Report*, 8 April. Available at: www.justice-report .com/bh/sadr%C5%BEaj-%C4%8Dlanci/vijest-stra%C5%A1na-no%C4%87 -u-viktor-bubnju (accessed 23 March 2015).

Van Marle K, de Villiers I and Beukes E (2012) Memory, Space and Gender: Re-imagining the Law. *Southern African Public Law* 27(2): 559–574.

Vercoe K (2013) *For Those Who Can Tell No Tales*. Paris: MPM Film.

Volcic Z, Erjavec K and Peak M (2013) Branding Post-war Sarajevo. Journalism, Memories, and Dark Tourism.*Journalism Studies* 15(6): 726–742.

Vujević A (2010 Zlocini u BiH: SarajevskamučenjaSrba [Crimes in BiH: Sarajevo Serbs Torture]. *Pecat*, 23 December. Available at: www.pecat.co.rs/2010/12/zloc ini-u-bih-sarajevska-mucenja-srba/ (accessed 23 March 2015).

Vulliamy E (2012) *The War Is Dead, Long Live the War: Bosnia the Reckoning*. London: Bodley Head.

Winstone M (2010) *The Holocaust Sites of Europe: An Historical Guide*. London: I.B. Tauris.

Živak S (1999) *Živim da svjedočim*. Beograd: Svet Knjige.

Index

Note: Locators followed by 'n' 't' and 'f' refer to notes, tables and figures respectively.

Printed in the United States
By Bookmasters